Fuel Cell Bus Life Cycle Cost Model: Base Case & Future Scenario Analysis

DOT-T-07-01   June, 2007

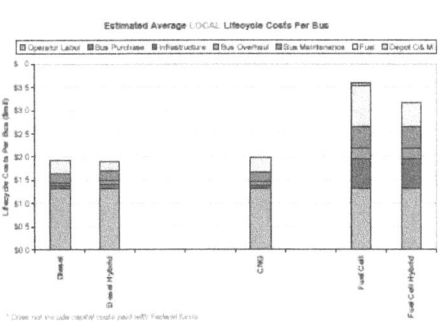

U.S. Department of Transportation

Research and Innovative Technology Administration

Office of Research Development and Technology

1200 New Jersey Ave. SE

Washington, DC 20590

REPORT DOCUMENTATION PAGE			Form Approved OMB No. 0704-0188

Public reporting burden for this collection of information is estimated to average 1 hour per response, including the time for reviewing instructions searching existing data sources, gathering and maintaining the data needed, and completing and reviewing the collection of information. Send comments regarding this burden estimate or any other aspect of this collection of information, including suggestions for reducing this burden, to Washington Headquarters Services Directorate for Information Operations and Reports, 1215 Jefferson Davis Highway Suite 1204, Arlington, VA 22202-4302, and to the Office of Management and Budget, Paperwork Reduction Project (0704-0188), Washington, DC 20503.

1. AGENCY USE ONLY (Leave blank)	2. REPORT DATE June 2007	3. REPORT TYPE AND DATES COVERED Final Report – June 2007	
4. TITLE AND SUBTITLE Fuel Cell Bus Life Cycle Cost Model: Base Case & Future Scenario Analysis		5. FUNDING NUMBERS	
6. AUTHOR(S) Dana Lowell (principle),* William P. Chernicoff,** F. Scott Lian***			
7. PERFORMING ORGANIZATION NAME(S) AND ADDRESS(ES) *MJ Bradley & Assoc., 1000 Elm St., 2nd Floor, Manchester, NH 03101 **US DOT-RITA, 1200 New Jersey Ave, SE, Washington, DC 20590 ***US DOT-Volpe National Transportation Systems Center, 55 Broadway, Cambridge, MA 02142		8. PERFORMING ORGANIZATION REPORT NUMBER DOT-T-07-01	
9. SPONSORING/MONITORING AGENCY NAME(S) AND ADDRESS(ES) U.S. Department of Transportation Research and Innovative Technology Administration Office of Research Development and Technology		10. SPONSORING/MONITORING AGENCY REPORT NUMBER DOT-T-07-01	
11. SUPPLEMENTARY NOTES None			
12a. DISTRIBUTION/AVAILABILITY STATEMENT This document is available to the public through the National Technical Information Service, Springfield, VA 22161		12b. DISTRIBUTION CODE	

13. ABSTRACT (Maximum 200 words)

This report describes the results of a life cycle cost analysis conducted using a spread sheet-based Lifecycle Cost Model developed to allow the user to evaluate the differential costs of different transit bus propulsion technologies. The model is set up to allow analysis of bus/technology types that operate on various liquid and gaseous fuels.

The model includes six input worksheets into which the user is required to enter various fleet data assumptions, and four output worksheets which display the costs calculated by the model for the bus/technology types analyzed.

The user can chose up to eight different bus/technology types at a time for analysis, organized by fuel type. The model allows simultaneous analysis of two different bus types operating on each of two different liquid fuels and two different bus types operating on each of two different gaseous fuels.

14. SUBJECT TERMS Hydrogen, fuel cell, cost analysis, life cycle analysis, financial forecasting, cost model			15. NUMBER OF PAGES	
			16. PRICE CODE	
17. SECURITY CLASSIFICATION OF REPORT Unclassified	18. SECURITY CLASSIFICATION OF THIS PAGE Unclassified	19. SECURITY CLASSIFICATION OF ABSTRACT Unclassified	20. LIMITATION OF ABSTRACT	

NSN 7540-01-280-5500

Standard Form 298 (Rev. 2-89) Prescribed by ANSI Std. 239 18 298-102

NOTICE

This document is disseminated under the sponsorship of the U.S. Department of Transportation in the interest of information exchange. The United States Government assumes no liability for its contents or use thereof.

NOTICE

The United States Government does not endorse products or manufacturers. Trade or manufacturers' names appear herein solely because they are considered essential to the objective of this report.

English to Metric Conversion Factors:

English to Metric	Metric to English
LENGTH (Approximate)	**LENGTH (Approximate)**
1 inch (in) = 2.5 centimeters (cm) 1 foot (ft) = 30 centimeters (cm) 1 yard (yd) = 0.9 meter (m) 1 mile (mi) = 1.6 kilometers (km)	1 millimeter (mm) = 0.04 inch (in) 1 centimeter (cm) = 0.4 inch (in) 1 meter (m) = 3.3 feet (ft) 1 meter (m) = 1.1 yards (yd) 1 kilometer (km) = 0.6 mile (mi)
AREA (Approximate)	**AREA (Approximate)**
1 square inch (sq in, in^2) = 6.5 square centimeters (cm^2) 1 square foot (sq ft, ft^2) = 0.09 square meter (m^2) 1 square yard (sq yd, yd^2) = 0.8 square meter (m^2) 1 square mile (sq mi, mi^2) = 2.6 square kilometers (km^2) 1 acre = 0.4 hectare (he) = 4,000 square meters (m^2)	1 square centimeter (cm^2) = 0.16 square inch (sq in, in^2) 1 square meter (m^2) = 1.2 square yards (sq yd, yd^2) 1 square kilometer (km^2) = 0.4 square mile (sq mi, mi^2) 10,000 square meters (m^2) = 1 hectare (he) = 2.5 acres
MASS-WEIGHT (Approximate)	**MASS-WEIGHT (Approximate)**
1 ounce (oz) = 28 grams (gm) 1 pound (lb) = 0.45 kilograms (kg) 1 short ton = 2,000 pounds (lb) = 0.9 tonne (t)	1 gram (gm) = 0.036 ounce (oz) 1 kilogram (kg) = 2.2 pounds (lb) 1 tonne (t) = 1,000 kilograms (kg) = 1.1 short tons
VOLUME (Approximate)	**VOLUME (Approximate)**
1 teaspoon (tsp) = 5 milliliters (ml) 1 tablespoon (tbsp) = 15 milliliters (ml) 1 fluid ounce (fl oz) = 30 milliliters (ml) 1 cup (c) = 0.24 liter (l) 1 pint (pt) = 0.47 liter (l) 1 quart (qt) = 0.96 liter (l) 1 gallon (gal) = 3.8 liters (l) 1 cubic foot (cu ft, ft^3) = 0.03 cubic meter (m^3) 1 cubic yard (cu yd, yd^3) = 0.76 cubic meter (m^3)	1 milliliter (ml) = 0.03 fluid ounce (fl oz) 1 liter (l) = 2.1 pints (pt) 1 liter (l) = 1.06 quarts (qt) 1 liter (l) = 0.26 gallon (gal) 1 cubic meter (m^3) = 36 cubic feet (cu ft, ft^3) 1 cubic meter (m^3) = 1.3 cubic yards (cu yd, yd^3)
TEMPERATURE (Exact)	**TEMPERATURE (Exact)**
$[(x - 32)(5/9)]\ °F = y\ °C$ $(x + 460)/1.8 = y\ °K$	$[(9/5)y + 32]\ °C = x\ °F$ $(y \times 1.8 - 460) = x\ °F$
PRESSURE (Exact)	**PRESSURE (Exact)**
1 psi = 6.8948 kPa	1 MPa = 145.04 psi
ENERGY & ENERGY DENSITY (Exact)	**ENERGY & ENERGY DENSITY (Exact)**
1 Btu = 1.05506 kJ 1 Btu/lb = 2.326 kJ/kg	1 MJ = 947.81 Btu 1 MJ/kg = 430 Btu/lb

QUICK FAHRENHEIT-CELSIUS TEMPERATURE CONVERSION

°F	-40°	-22°	-4°	14°	32°	50°	68°	86°	104°	122°	140°	158°	176°	194°	212°
°C	-40°	-30°	-20°	-10°	0°	10°	20°	30°	40°	50°	60°	70°	80°	90°	100°

Fuel Cell Bus Life Cycle Cost Model: Best Case & Future Scenario Analysis

Table of Contents

EXECUTIVE SUMMARY 1

1. LIFE CYCLE COST BASE CASE ASSUMPTIONS 4
   1.1 VEHICLES/TECHNOLOGIES AND FUELS ANALYZED 4
   1.2 DATA INPUTS 5
       1.2.1 Depot Baseline Data (Worksheet I1) 5
       1.2.2 Annual Bus Costs (Worksheet I2) 7
       1.2.3 Bus Purchase & Overhaul Costs (Worksheet I3) 13
       1.2.4 Variable Overhaul Intervals (Worksheet I4) 14
       1.2.5 Depot Infrastructure Costs (Worksheet I5) 15
       1.2.6 Bus Technology Training Requirements (Worksheet I6) 18

2. BASE CASE RESULTS 19
   2.1 FIRST YEAR ANNUAL COSTS (WORKSHEET O1) 19
   2.2 CAPITAL COSTS (WORKSHEET O2) 20
   2.3 OVERHAUL COSTS PER BUS (WORKSHEET O3) 21
   2.4 TOTAL LIFE CYCLE COSTS (WORKSHEET O4) 22

3. FUTURE COST SCENARIOS 27
   3.1 "BEST CASE" ASSUMPTIONS 27
   3.2 "BEST CASE" RESULTS 28
   3.3 SENSITIVITY ANALYSIS – CAPITAL AND FUEL COSTS 30

APPENDIX A – Base Case Life Cycle Cost Model Input & Output Sheets

Executive Summary

This report describes the results of a life cycle cost analysis conducted using a spread sheet-based Lifecycle Cost Model developed to allow the user to evaluate the differential costs of different transit bus propulsion technologies. The model is set up to allow analysis of bus/technology types that operate on various liquid and gaseous fuels[1].

The model includes six input worksheets into which the user is required to enter various fleet data assumptions, and four output worksheets which display the costs calculated by the model for the bus/technology types analyzed.

The user can chose up to eight different bus/technology types at a time for analysis, organized by fuel type. The model allows simultaneous analysis of two different bus types operating on each of two different liquid fuels and two different bus types operating on each of two different gaseous fuels. The five fuel/technology combinations analyzed and presented here are shown in Table 1.

Table 1 Fuel/Technology Combinations Analyzed

Fuel	Propulsion Technology
Liquid Fuel 1 – Standard Diesel Fuel (ULSD)	Standard Diesel Propulsion (Diesel)
	Diesel Hybrid Electric Propulsion (Diesel Hybrid)
Liquid Fuel 2 – NONE	None
Gaseous Fuel 1 – Compressed Natural Gas (CNG)	Standard Natural Gas Propulsion (CNG)
Gaseous Fuel 2 – Compressed Hydrogen (C-H_2)	Fuel Cell Electric Drive (Fuel Cell)
	Fuel Cell Hybrid Drive (Fuel Cell Hybrid)

These fuel/technology combinations were chosen to be illustrative of currently available and developing technologies, and to demonstrate the utility of the life cycle cost model used. These fuel/technologies combinations do not represent the only ones that could have been analyzed.

For all fuel and technology combinations the base vehicle is assumed to be a new 40-foot low-floor urban transit bus. The analysis assumes that all bus sub-systems other than the power plant, drive system, and fuel system (e.g. brakes, suspension, air conditioning, customer amenities, etc.) are identical on all of the bus types analyzed.

Elements of total life cycle cost included in the analysis include the following capital and annual operating costs:

---

[1] This model is documented in the report *Fuel Cell Bus Life Cycle Cost Model*, May 2007, prepared by M.J. Bradley & Associates for the Volpe National Transportation Systems Center.

Fuel Cell Bus Life Cycle Cost Model: Best Case & Future Scenario Analysis

CAPITAL COSTS

- bus purchase
- purchase/installation of required fueling infrastructure
- purchase/installation of required depot modifications, special tools, and special infrastructure
- initial operator, mechanic and manager training;

ANNUAL OPERATING COSTS

- annual operator labor costs
- annual bus maintenance costs
- annual bus fuel costs
- annual maintenance and operating cost of required fueling infrastructure, depot modifications, special tools, and special infrastructure
- periodic bus overhaul costs
- annual refresher training costs.

The "base case" analysis is intended to evaluate current costs for fuel cell buses compared to other technology options, recognizing that fuel cells are still an emerging technology while the other analyzed options are more mature. Many of the cost assumptions used in the base case analysis are based on data reported by the National Renewable Energy Laboratory's (NREL) Advanced Vehicle Testing Activity. Seven NREL reports were reviewed, which covered three small-scale fuel cell bus demonstration deployments, two diesel hybrid-electric bus deployments, and two natural gas bus deployments. Other assumptions are based on data reported in the Federal Transit Administration's National Transit Database, and discussions with vehicle and technology manufacturers and transit maintenance managers.

The base case analysis shows that current total capital costs, first year annual costs, average annual costs, and total life cycle costs are significantly higher for a fleet of 100 Fuel Cell or Fuel Cell Hybrid buses than for a 100-bus fleet of Diesel, CNG, or Diesel Hybrid buses. The net present value of projected total life cycle costs averages approximately $6 million per bus for Fuel Cell and Fuel Cell Hybrid buses compared to $2 million per bus for Diesel, CNG, and Diesel Hybrid buses. Projected average total per-mile costs for Fuel Cell buses are $15.78/mile and for Fuel Cell Hybrid buses are $14.70/mile, compared to $5.58 - $5.90/mile for Diesel, CNG, and Diesel Hybrid buses.

The single largest contributor to the increased life cycle costs for Fuel Cell and Fuel Cell Hybrid buses is the increased capital cost to purchase buses and install necessary infrastructure. However, all cost elements other than operator labor costs are significantly higher for fuel cell buses than for the other bus types, including life time overhaul costs (~3x higher), annual maintenance costs (~2 x higher), and fuel costs (~3x higher for Fuel Cell and ~2x higher for Fuel Cell Hybrid).

If only local costs are included, by removing the portion of capital costs paid with federal funds, average per-mile life cycle costs for Fuel Cell and Fuel Cell Hybrid buses fall to

$9.15/mile and $8.10/mile, respectively. These per-mile local costs are still 60-90% higher than local per-mile costs for operation of diesel buses.

Operator costs make up approximately 60% of current total life cycle costs for Diesel, CNG, and Diesel Hybrid buses; the second largest cost element is amortization of capital costs, at approximately 15%. With Fuel Cell buses amortization of capital costs accounts for over 50% of total life cycle costs, pushing operator costs down to only 21% of the total. Though higher in absolute value for Fuel Cell buses than for the other bus types the other cost categories (overhaul costs, maintenance costs, fuel costs, and depot costs) comprise a similar percentage of the total.

If only local costs are included, operator labor accounts for over 68% of total costs for diesel buses, while fuel accounts for over 14% of costs and capital amortization only accounts for a little over 3% of costs. By contrast operator labor only accounts for about 36% of local costs for Fuel Cell Buses while fuel accounts for 25% of local costs and capital amortization accounts for almost 18% of local costs.

The life cycle cost model was also used to conduct a near-term "best case" analysis, which is based on meeting the Federal Transit Administration's National Fuel Cell Bus performance objectives, and the U.S. Department of Energy's 2015 goal for the cost of hydrogen fuel. These goals include a 50% reduction in the purchase price of fuel cell buses, a doubling of fuel cell stack life, a significant improvement in fuel economy, and greater than 50% reduction in the cost of hydrogen fuel compared to the base case. To meet the FTA fuel economy targets it was assumed that any fuel cell bus would have to use a hybrid electric propulsion system.

Under the best case scenario, total per-mile life cycle costs for Fuel Cell Hybrid buses fall by 40% compared to the base case, to $8.88/mile. If only local costs are included best case average per-mile life cycle costs for Fuel Cell Hybrid buses fall to $5.49/mile - $0.58/mile more than local life cycle costs for Diesel buses.

Under the best case scenario the single largest contributor to higher life cycle costs for Fuel Cell Hybrid buses is still capital amortization due to a higher bus purchase price and higher infrastructure costs for hydrogen fueling. Under the best case scenario capital amortization accounts for almost 48% of total life cycle costs for Fuel Cell Hybrid buses, compared to 15% for diesel buses. With all other best case assumptions held constant, a Fuel Cell Hybrid bus would have to cost no more than $350,000 (less than the price of current CNG buses) for total life cycle costs to fall to the level of costs for Diesel buses. In order to match local life cycle costs for Diesel buses a Fuel Cell Hybrid bus could cost no more than $500,000 (approximately the current price of diesel hybrid buses).

Under the best case scenario life cycle fuel costs for Fuel Cell Hybrid buses are significantly lower than for Diesel buses, partially off-setting increased life cycle costs for capital amortization, maintenance, and overhauls. The lower the price of hydrogen fuel, the greater the reduction. However, the life cycle cost model shows that even if hydrogen fuel were free the fuel cost savings from Fuel Cell Hybrid buses would not fully off-set the increases in other cost categories compared to diesel buses.

1. Life Cycle Cost Base Case Assumptions

This section describes the fuel/technology combinations analyzed and the major cost assumptions used in the base case analysis for each; the sources of all major assumptions are noted.

Many of the cost assumptions used in this analysis are based on data reported by the National Renewable Energy Laboratory's (NREL) Advanced Vehicle Testing Activity. Seven recent NREL reports were reviewed, which covered three fuel cell bus deployments, two diesel hybrid-electric bus deployments, and two natural gas bus deployments. Other assumptions are based on data reported in the Federal Transit Administration's National Transit Database, and discussions with vehicle and technology manufacturers and transit maintenance managers.

1.1 Vehicles/Technologies and Fuels Analyzed

The five fuel/technology combinations analyzed here represent the most common existing and emerging options for powering U.S. transit buses. Currently approximately 82% of U.S. transit buses are powered by diesel engines and 15% are powered by natural gas engines[2]. Hybrid-electric drive is also growing in popularity as an alternative to standard propulsion for buses, with over 1,600 diesel hybrid buses in service in 2007 and almost 900 more on order[3].

Fuel cells are an emerging technology for buses. To date only small scale demonstration fleets have been put into service, and there are currently eight fuel cell transit buses operating in California and Connecticut[4].

The five fuel/technology combinations chosen for analysis do not represent the only options currently in service or under development. They were chosen to be illustrative of available options and to demonstrate the utility of the life cycle cost model used. Other fuel/technology combinations that could have been analyzed using the model include gasoline hybrid-electric propulsion, and internal combustion engines operating on hydrogen fuel.

Table 2 shows the major elements of the propulsion system assumed to be included on each of the bus types analyzed. All other bus systems are assumed to be identical.

Both the Diesel and Diesel Hybrid buses are assumed to operate on standard on-highway diesel fuel, which since late 2006 has been "ultra-low sulfur diesel" (ULSD) with less than 15 parts per million sulfur.

CNG buses are assumed to operate on natural gas which is delivered to and stored on the vehicle in compressed form at maximum pressures of 3,600 pounds per square inch (standard in the transit industry).

---

[2] American Public Transportation Association. 2006 survey data. <http://www.apta.com/research/stats/bus/power.cfm>

[3] 2006 APTA survey and discussion with bus manufacturers.

[4] These buses are operated by the Alameda Contra Costa Transit District (3), the Santa Clara Valley Transportation Authority, the Sunline Transit Agency (1), and Connecticut Transit (1)

The engines used in the Diesel, Diesel Hybrid, and CNG buses are assumed to be compliant with 2007 EPA emissions standards for new heavy-heavy duty engines.

Both Fuel Cell and Fuel Cell Hybrid buses are assumed to operate on hydrogen gas which is delivered to and stored on the vehicle in compressed form at maximum pressures of 5,000 pounds per square inch (standard for current fuel cell buses).

Table 2 Propulsion System Components

Bus Type	Powerplant	Drive System	Fuel System
Diesel	Compression ignition internal combustion engine (diesel)	5-speed automatic transmission	Diesel fuel storage system
Diesel Hybrid	Compression ignition internal combustion engine (diesel)	Series hybrid drive system[5] • traction generator • electric traction motor • energy storage system • inverter/power electronics	Diesel fuel storage system
CNG	Spark ignition internal combustion engine (natural gas)	5-speed automatic transmission	Compressed natural gas storage system (3,600 psi)
Fuel Cell	Proton exchange membrane fuel cell engine	Electric drive system • electric traction motor • inverter/power electronics	Compressed hydrogen gas storage system (5,000 psi)
Fuel Cell Hybrid	Proton exchange membrane fuel cell engine	Series hybrid drive system • electric traction motor • energy storage system • inverter/power electronics	Compressed hydrogen gas storage system (5,000 psi)

1.2 Data Inputs

The following describes the sources of the major cost assumptions used in the analysis for each fuel/technology combination.

1.2.1 Depot Baseline Data (Worksheet I1)

For this analysis buses are assumed to be assigned to a notional 100-bus depot facility, which is a typical size for many U.S. transit operations. To maximize necessary depot

---

[5] Series hybrid technology was chosen to provide a more direct comparison to electric drive systems used in Fuel Cell and Fuel Cell Hybrid buses. Parallel hybrid drive systems are also commercially available for transit buses.

and fueling investments it is assumed that all buses assigned to the depot will be of the same type.

Depot personnel assignments for a 100-bus depot are assumed to be as follows:

- Bus operators – 300 (assuming 24-hr operations and 85% employee availability)
- Bus mechanics – 20 (consistent with maintenance cost assumptions noted below)
- Managers – 30 (one manager, including foremen, for every ten hourly employees)

Note that in the model these personnel assignment numbers are only used to calculate training costs.

Bus mechanics are assumed to have a fully-loaded labor rate of $50/hour. This is consistent with the data used to determine average bus maintenance costs, as discussed in Section 1.2.2 below. Bus operators are also assumed to have a fully-loaded labor rate of $50/hour and managers are assumed to have a fully-loaded labor rate of $75/hour.

The assumptions used in this analysis for diesel fuel and natural gas commodity costs were taken from the U.S. Department of Energy's Clean Cities Alternative Fuel Price Report for March 2007. That report shows that in March 2007 the average price of diesel fuel at 333 public gas stations surveyed was $2.63/gallon (and it ranged from an average of $2.48/gallon on the Gulf Coast to $2.96/gallon on the West Coast). Compressed natural gas was also sold at 123 of the same stations, and it's price averaged $2.17/diesel-equivalent gallon (ranging from $1.56/DEG in the mid-west to $2.83/DEG in New England).

Three of four U.S. transit agencies currently operating fuel cell buses report that the cost of producing and delivering compressed hydrogen to their buses ranges from $4.26/kg to $9.06/kg (see Table 4 below). This is equivalent to $4.81 - $10.23/DEG[6]. This analysis assumes that compressed hydrogen will cost $6.70/kg, or $7.57/DEG.

Capital Cost Share is assumed to be 80% for the federal government and 20% for a local match. This is typical for capital funding provided by the Federal Transit Administration.

Annual inflation is assumed to be 2.3% for fuel and 2.3% for labor and materials (including bus overhaul costs). This is in line with current market expectations for long-term inflation, as calculated by the difference in the yields of long-term nominal U.S. treasury notes and treasury inflation-protected securities (TIPS)[7].

A 5% discount rate is used for net-present-value calculations. This includes the expected inflation noted above plus a 2.7% "real discount rate" to account for risk return on invested capital. This risk return value is equivalent to the current rate of return on treasury inflation-protected securities [8].

---

[6] Assuming 128,400 btu/gallon for diesel and 113,628 btu/kg for hydrogen = 1.13 kg/diesel gallon.

[7] See information from the Federal Reserve Bank of Cleveland <http://www.clevelandfed.org/research/inflation/TIPS/index.cfm>

[8] See Daily Treasury Real Long Term rates as calculated by the U.S. Treasury. <http://www.ustreas.gov/offices/domestic-finance/debt-management/interest-rate/real_yield_historical.shtml>

The analysis also assumes that no programmed overhauls will be performed within two years of retirement of any bus. This precludes the model from assuming that a major investment will be made in any bus just prior to retirement.

1.2.2 Annual Bus Costs (Worksheet I2)

In this analysis the useful life for all buses is assumed to be 12 years. This is the minimum in-service age at which transit agencies which use federal funds for bus purchase can retire buses, per FTA rules, and is a standard widely used in the transit industry for planning and financial analysis.

To determine appropriate assumptions for annual mileage per bus, and average in-service speed, data on bus operations reported to the National Transit Database[9] was analyzed. This data is summarized in Table 3. As shown, for over 42,000 buses operated by 374 U.S. transit agencies the average in-service speed in 2005 was 12.4 mph, and the average annual mileage was 32,602 miles per bus. These assumptions were used in the analysis for all bus types.

Table 3 Summary - 2005 National Transit Database - Bus Mode

Fuel Use	Agencies	Buses [1]	Annual Miles/bus [2]			Average Speed			MPG [3]		
			low	AVG	*high*	*low*	AVG	*high*	*low*	AVG	*high*
> 75% Diesel	334	34,503	7,084	32,096	70,225	7.9	12.4	50.1	2.1	3.2	9.8
Diesel - CNG Mixed	18	3,290	21,661	35,220	43,703	10.2	12.5	17.9	2.0	2.9	4.3
>75% NG	22	4,391	18,679	34,620	54,266	9.6	12.2	21.4	1.8	2.4	4.1
TOTAL	374	42,184		32,602			12.4			3.0	

[1] Reported Vehicles Operated in Maximum Service (VOMS)
[2] Based on VOMS plus 15% spares
[3] Miles per diesel equivalent gallon

Assumptions about average fuel economy for Diesel and CNG buses were also taken from the NTD data. As shown in Table 3 predominantly diesel fleets (>75% of reported fuel use diesel) report significantly higher average fuel economy than predominantly CNG fleets (>75% of reported fuel use NG) - 3.2 MPG versus 2.4 MPG. The analysis used these average values for Diesel and CNG bus fuel economy. High and low values were entered as +/- 20% of these averages, to account for variability from fleet to fleet. For both predominantly diesel and predominantly NG fleets in the NTD database, average fuel economy data covering approximately 80% of reported buses is within +/- 20% of the total fleet average. These assumptions are also in agreement with data reported by NREL for operations with similar average speed (~12 mph) – see Tables 4 and 5 below.

The model calculates basic annual bus maintenance costs based on $/mile cost factors for propulsion system-related and non-propulsion-related maintenance. To determine appropriate assumptions for these maintenance cost factors, and for Hybrid and Fuel Cell

---

[9] Federal Transit Administration, 2005 National Transit Database, Tables 17 and 19.
<http://www.ntdprogram.com/ntdprogram/pubs.htm>

bus average fuel economy, seven NREL bus evaluation reports were reviewed. The data from these reports is summarized in Tables 4 and 5.

As shown in these tables non-propulsion related maintenance costs for most of the buses covered by these analyses ranged from $0.23 - $0.54/mile[10]. For this analysis we assumed that all buses would have non-propulsion related maintenance costs of $0.40/mile +/- $0.15/mile.

Table 4   Summary of Results from NREL Fuel Cell Bus Evaluations

		Unit	AC TRANSIT 4/06 - 11/06 40-ft Fuel Cell Hybrid 40-ft Diesel		VTA 11/04 - 7/06 40-ft Fuel Cell 40-ft Diesel		SUNLINE 1/06 - 11/06 40-ft Fuel Cell Hybrid 40-ft CNG	
Capital ($ mill)	H2 Fuel Station Installation	total	not reported		$ 0.64		not reported	
	H2 Depot Modifications	total	$ 1.50		$ 4.40		$ 0.05	
	Fuel Cell Bus Purchase	ea	$ 3.20		$ 3.50		$ 3.10	
Fuel Economy	Duty Cycle	MPH	11.6		14.5		13.0	
	Fuel Cell Fuel Economy	mi/kg	5.50		3.12		7.3	
		MPDEG	6.22		3.52		8.28	
	Diesel Fuel Economy	MPG	4.00		3.98		CNG = 3.32	
Fuel Cost	Hydrogen Cost	$/kg	$ 8.00		$ 9.06		$ 4.26	
	Diesel Cost	$/gal	$ 2.30		$ 2.07		CNG = $1.10	
			Fuel Cell	Diesel	Fuel Cell	Diesel	Fuel Cell	CNG
Maintenance Cost	PMI	$/mi	$ 0.15	$ 0.08	$ 0.61	$ 0.09	$ 0.05	$ 0.08
	Powerplant	$/mi	$ 0.01	$ 0.10	$ 1.54	$ 0.16	$ 0.11	$ 0.05
	Drive System	$/mi	$ 0.04	$ -	$ 0.36	$ 0.02	$ 0.06	$ -
	Fuel System	$/mi	$ 0.01	$ 0.02	$ 0.48	$ 0.02	$ -	$ 0.01
	TOTAL PROPULSION	$/mi	$ 0.06	$ 0.12	$ 2.38	$ 0.20	$ 0.17	$ 0.06
	NON-PROPULSION	$/mi	$ 0.54	$ 0.23	$ 1.17	$ 0.34	$ 0.27	$ 0.19
	TOTAL	$/mi	$ 0.60	$ 0.35	$ 3.55	$ 0.54	$ 0.44	$ 0.25
NOTES			Fuel cell maintenance done under warranty - costs not included. Fuel cell bus + 8,000 lb Van Hool buses. Fuel cell buses use ISE drive system with ZEBRA batteries		Warranty parts costs not included above: $13.29/mi fuel cell; $0.04/mi diesel. Fuel cell bus + 6,800 lb. Fuel station includes one dispenser and 10 min fill; designed for 6 buses. Fuel station vent and boil off losses ~50%. New Flyer buses. Fuel cell buses use Ballard drive system		Fuel cell maintenance done under warranty - costs not included. Fuel cell bus + 8,000 lb Van Hool buses. Fuel cell buses use ISE drive system with ZEBRA batteries. H2 maintenance facility is a tent - two bays	
			For all agencies maintenance costs calculated using $50/hr mechanic labor rate					
SOURCES			AC Transit NREL/TP-560-41041 March 2007 VTA NREL/TP-560-40615 November 2006 Sunline NREL/TP-560-41001 February 2007					

With the exception of both CNG and hybrid buses at NYCT total propulsion-related maintenance costs for diesel, natural gas, and hybrid buses in these studies ranged from $0.06 - $0.20/mile. A direct comparison of natural gas and hybrid bus costs to diesel bus costs at the same agency indicates that both natural gas and hybrid buses have the same, or only marginally higher, propulsion-related maintenance costs as diesel buses. For this

---

[10] The exceptions were both hybrid and CNG buses at NYCT – whose costs were similar, but higher than at other agencies – and fuel cell buses at VTA, which had significantly higher costs than the comparison diesel buses.

study we assumed that diesel buses have propulsion-related maintenance costs of $0.15/mile +/- $0.05/mile. Both CNG and Hybrid buses were assumed to have propulsion-related maintenance costs $0.01/mile higher than diesel buses.

Propulsion-related maintenance costs reported by NREL for fuel cell buses were much more variable. At AC Transit reported $/mile costs for propulsion-related maintenance were actually lower for the fuel cell buses than for the comparison diesel buses, while at Sunline they were almost three times higher, and at VTA they were almost 12 times higher ($2.38/mile).

Table 5 Summary of Results from NREL Hybrid and Natural Gas Bus Evaluations

		Unit	DART 6/98 - 1/00 40-ft LNG 40-ft Diesel		WMATA 9/01 - 9/04 40-ft CNG 40-ft Diesel		KC METRO 4/05 - 3/06 60-ft Diesel Hybrid 60-ft Diesel		NYCT 10/04 - 8/05 40-ft Diesel Hybrid 40-ft CNG	
Capital ($ mill)	Diesel Fuel Station	total	not reported		not reported		not reported		not reported	
	NG Fuel Station	total	$ 7.50		$ 4.00		NA		$ 7.40	
	Hybrid Depot Modifications	total	NA		NA		None		/depot - 2 battery conditi	
	NG Depot Modifications	total	incl in NG fuel station		$ 11.60		NA		not reported	
	Hybrid Bus Purchase	ea	NA		NA		$ 0.645		not reported	
	NG Bus Purchase	ea	not reported		$ 0.34		NA		not reported	
Fuel Economy	Duty Cycle	MPH	13.7 - 14.4		11.6		11.6 - 12.4		6.2 - 6.5	
	Hybrid Fuel Economy	MPG	NA		NA		3.17		3.2	
	NG Fuel Economy	MPDEG	2.70		2.32 - 2.39		NA		1.7	
	Diesel Fuel Economy	MPG	3.80		2.84		2.50		2.30 - 2.40	
Fuel Cost	NG Cost	$/DEG	$ 0.82		$ 1.50		NA		$ 1.74	
	Diesel Cost	$/gal	$ 0.90		$ 1.33		$ 1.98		$ 1.78	
			LNG	Diesel	CNG	Diesel	Hybrid	Diesel	Hybrid	CNG
Maintenance Cost	PMI	$/mi	$ 0.07	$ 0.07	$0.12-$0.14	$ 0.17	$ 0.05	$ 0.05	$ 0.17	$ 0.12
	Powerplant	$/mi	$ 0.08	$ 0.06	$0.11-$0.12	$ 0.11	$ 0.11	$ 0.12	$ 0.17	$ 0.25
	Drive System	$/mi	$ 0.02	$ 0.01	$0.01-$0.03	$ 0.04	$ 0.01	$ -	$ 0.18	$ 0.04
	Fuel System	$/mi	$ 0.01	$ 0.01	$0.01-$0.02	$ 0.01	$ 0.01	$ -	$ 0.02	$ 0.06
	TOTAL PROPULSION	$/mi	$ 0.11	$ 0.08	$0.13-$0.17	$ 0.16	$ 0.13	$ 0.12	$ 0.37	$ 0.35
	NON-PROPULSION	$/mi	$ 0.29	$ 0.45	$0.39-$0.41	$ 0.43	$ 0.31	$ 0.34	$ 0.86	$ 0.94
	TOTAL	$/mi	$ 0.40	$ 0.53	$0.52-$0.58	$ 0.59	$ 0.44	$ 0.46	$ 1.23	$ 1.29
NOTES			Capital costs are for two LNG fuel stations and modifications at two depots. Warranty costs not included		CNG cost includes $0.14/DEG for compression (electricity) and $0.17/DEG for fuel station O&M ($360,000/yr). Warranty costs not included		All buses 60' New Flyer D60LF. Hybrids included Allison EP50™ system. Warranty costs of $0.17 - $0.20/mi for both diesel and hybrid buses. Diesel and hybrid buses operated from different depots. A direct route comparison showed a 21-22% increase in FE for hybrids		CNG fuel station is 6,600 scfm. Cost includes $2 mill for NG line extension to depot. CNG cost includes $0.35/DEG for compression and fuel station maintenance. All buses were 40' Orion VII. Hybrids included BAE HybriDrive™ system. Warranty costs not included	
			For all agencies maintenance costs calculated using $50/hr mechanic labor rate							
SOURCES			DART NREL, Dart's LNG Bus Fleet Final Results, October 2000 WMATA NREL/TP-540-37626 April 2006 KC Metro NREL/TP-540-40585 December 2006 NYCT NREL/TP-540-40125 November 2006							

At both AC Transit and Sunline virtually all propulsion-related maintenance during the study period was done by the manufacturer under warranty and is not included in the reported costs. VTA took greater responsibility for fuel cell bus maintenance and their reported costs are likely more representative. Based on availability and reliability statistics for the AC Transit and Sunline fuel cell buses it is clear that they too required significantly more maintenance than the comparison diesel buses during the study period.

Despite requiring more maintenance the actual $/mile costs reported for VTA fuel cell buses are somewhat misleading because these buses only accumulated one fifth the mileage of the comparison diesel buses during the study period. For this analysis we used a conservative, forward-looking assumption of $1.00/mile +/- $0.25/mile for propulsion-related maintenance costs for both Fuel Cell and Fuel Cell Hybrid buses.

Assumptions about Diesel Hybrid, Fuel Cell, and Fuel Cell Hybrid fuel economy were also taken from the NREL data. As shown in Table 5 the Diesel Hybrid buses operated by KC Metro had 21 – 27% better fuel economy than the comparison diesel buses, on a duty cycle very similar to the one chosen for this analysis (~12.4 mph). The Diesel Hybrid buses operated by NYCT had even higher relative fuel economy (36% better than diesel and 88% better than CNG), but on a much slower duty cycle (6.2 – 6.5 mph) which is advantageous to hybrid buses. For this analysis we assumed that Diesel Hybrid buses will have 25% better fuel economy than Diesel buses.

As shown in Table 4 the Fuel Cell buses operated by VTA had 12% worse fuel economy than the comparison diesel buses (miles per diesel equivalent gallon, MPDEG); this is the assumption that was used for this analysis. As shown in Table 4 the Fuel Cell Hybrid buses operated by AC Transit had 55% better fuel economy (MPDEG) than the comparison diesel buses and the Fuel Cell Hybrid buses operated by Sunline had 149% better fuel economy than the comparison CNG buses. This analysis assumes that Fuel Cell Hybrid buses will get 60% better fuel economy than diesel buses and 112% better fuel economy than CNG buses. The fuel economy assumptions used in the analysis for all bus types are shown in Table 6.

Table 6 Fuel Economy Assumptions Used in the Analysis

Bus Type	Fuel Economy, Miles per Diesel Equivalent Gallon		
	Low	AVG	High
Diesel	2.6	3.2	3.8
Diesel Hybrid	3.3	4.0	4.8
CNG	1.9	2.4	2.9
Fuel Cell	2.3	2.8	3.3
Fuel Cell Hybrid	4.2	5.1	6.1

The model calculates the cost of brake relines separately from base $/mile maintenance costs because hybrid propulsion systems have been shown to significantly extend brake reline intervals due to regenerative braking. In addition, CNG and Fuel cell buses are typically up to 25% heavier than diesel buses due to the greater weight of the gaseous fuel system and other components, which reduces reline intervals since the braking system needs to do more work to stop the bus.

Table 7 contains the values used in the analysis for front and rear reline interval, front and rear reline material cost, and front and rear reline labor hours for Diesel buses. These assumptions are based on an informal poll of maintenance staff at six transit agencies conducted by the author in 2004[11]. For all other bus types the brake reline material costs and labor hours are assumed to be the same as for Diesel buses.

For CNG buses brake reline intervals are assumed to be 10% shorter (worse) than for Diesels due to the greater bus weight. For Fuel Cell buses brake reline intervals are assumed to be 15% shorter.

Given that significant numbers of hybrid buses have not been in service for more than a few years, hard data on brake life does not yet exist. However, anecdotal evidence from several maintenance managers with hybrid experience indicates that brake lining life on hybrids may be more than double brake lining life on conventional buses. This is consistent with in-use fuel economy results for hybrids. A 20% reduction in fuel use for a hybrid bus implies that the braking system is recapturing about half the energy normally dissipated in braking, and that therefore the braking system is only doing about half the work that it would on a conventional bus[12], which implies that the bus should only require relines half as often. This analysis uses a conservative assumption of a 75% increase in reline interval for Diesel Hybrid buses and a 60% increase in reline interval for Fuel Cell Hybrid buses (the difference is due to the greater weight of fuel cell hybrids).

	Unit	Value
Front Interval	mi	35,000
Rear Interval	mi	30,000
Front Matl Cost	$	$400
Rear Matl Cost	$	$400
Front Labor	hr	5
Rear labor	hr	8

Table 7 Brake Maintenance Assumptions, Diesel Buses

The model also allows a user to specify up to five different "technology-specific" maintenance costs, over and above base propulsion-related costs, in order to better evaluate the differences

---

[11] The agencies polled included: Dallas Area Rapid Transit, Dallas, TX, Toronto Transit Commission, Toronto, ON, Washington Metropolitan Area Transit Authority, Washington, DC, MTA New York City Transit, Brooklyn, NY, Coast Mountain Bus Company, Vancouver, BC, Los Angeles County Metropolitan Transportation Authority, Los Angeles, CA.

[12] On a typical transit bus approximately 20% of the energy supplied by the engine is used to operate accessory loads, and 80% is supplied to the bus wheels. Of the energy supplied to the bus wheels, approximately one half (40% of the total) is dissipated as friction between the tires and the road, and half (40% of total) is dissipated in the brake system. Assuming that all of the fuel savings from a hybrid bus comes from energy recovered through regenerative braking, a 20% savings implies that the brake system in only dissipating half the energy that it would on a standard bus.

between technologies. In this analysis only one technology-specific maintenance item was included - diesel particulate filter cleaning - which is applicable to Diesel and Diesel Hybrid buses.

Diesel particulate filters (DPF) are required on all 2007 model year and later diesel engines, to reduce emissions of particulate matter. DPFs must be removed periodically to have accumulated ash removed. This ash accumulates as engine lubricating oil is burned in the cylinder, since inorganic components of the oil can not oxidize out of the filter along with collected carbon. The actual cleaning interval will depend on duty cycle and how much oil the engine burns. However, most filter manufacturers recommend a base cleaning interval of once per year. This annual interval is the assumption used in this analysis.

Based on the author's experience at New York City Transit, the cost of this annual cleaning is $300 to $400 per bus. This includes two hours for removal/replacement of the DPF and a third-party cleaning fee of $200 - $300 per DPF. The model applies this annual DPF cleaning cost to Diesel buses and Diesel Hybrid buses.

All hybrid-electric propulsion systems use an energy storage sub-system to act as a load leveler during vehicle operation (supplying peak electrical power and absorbing electrical power during braking). There are a number of different energy storage technologies commercially available, including lead-acid batteries, nickel-metal hydride batteries, sodium/nickel chloride batteries, lithium ion batteries, and ultra-capacitors. Different manufacturers have made different commercial decisions about which battery technology to supply with their hybrid drive systems[13]. Some battery technologies require periodic maintenance, while others do not[14]. To provide a consistent comparison this analysis assumes that both Diesel Hybrid and Fuel Cell Hybrid buses will be equipped with either nickel-metal hydride or lithium-ion batteries, neither of which require regular maintenance. It is the author's judgment, based on current commercial developments, that these are the most likely energy storage technologies to be used for future hybrid bus deliveries in 2008 and beyond.

Operator labor rates were assumed to be $50/hr for all bus types, equivalent to labor rates for bus mechanics.

---

[13] The three leading U.S. heavy-duty drive system suppliers all use different technologies. BAE Systems Controls currently supplies commercial hybrid systems with lead-acid battery packs, but recently announced that they would switch to lithium-ion batteries beginning in 2008. Allison Electric Drives supplies commercial systems with nickel-metal hydride battery packs, while ISE has recently supplied systems using both ultra-capacitors and sodium/nickel chloride batteries.

[14] Lead-acid batteries used in a hybrid system typically require twice-yearly "conditioning" charging to reverse negative plate sulfation. Sodium/nickel chloride batteries operate at approximately 260°C, and often must be plugged into grid electrical power to maintain this temperature if the bus will not be used for an extended period. The other battery technologies do not require regular maintenance or charging in a hybrid application.

1.2.3 Bus Purchase & Overhaul Costs (Worksheet I3)

To determine average vehicle purchase costs for Diesel, CNG, and Diesel Hybrid buses data was gathered from the American Public Transportation Association 2006 Transit Vehicle Database[15]. Table 9 summarizes this data on the weighted average price for 35-ft and 40-foot buses purchased for delivery in 2005 and 2006. The 2006 values for 40-ft buses were used in the analysis for the purchase cost of Diesel, CNG, and Diesel Hybrid buses.

Table 8 Weighted Average Bus Purchase Prices (2006 APTA Transit Vehicle Database)

Year		35 Ft Buses			40 Ft Buses			% Diff NG	% Diff HYB
		Diesel	NG	% Diff	Diesel	NG	D-Hybrid		
2005	Price	$276,487	NA	NA	$329,076	$358,673	$541,281	9%	64%
	Num	231	0		991	463	183		
2006	Price	$277,357	$331,001	19%	$327,450	$376,667	$502,082	15%	53%
	Num	62	14		1,030	54	86		

In this analysis both Fuel Cell and Fuel Cell Hybrid buses are assumed to cost $3.2 million each. This is consistent with pricing reported by NREL for the three most recent fuel cell bus deliveries (see Table 4).

In order to maintain their buses in service for twelve years or more most transit agencies regularly overhaul them. The life cycle cost model used for this analysis allows the user to separately specify overhaul costs and overhaul intervals (in miles or hours of operation) for the following six bus sub-systems:

- Engine/power plant overhaul
- Transmission/drive system overhaul
- Bus overhaul (non-propulsion related systems)
- Technology Specific overhaul A
- Technology Specific overhaul B
- Technology Specific overhaul C

The technology-specific overhaul categories are designed to allow the user to separately identify items such as hybrid battery system replacements, which is the only technology specific overhaul category used in this analysis.

For all bus types the analysis assumes that a Bus Overhaul will happen at 200,000 miles (6 years, or mid-life of the bus) and cost $50,000. Table 9 contains the values used in this analysis for the cost and interval of engine/powerplant and transmission/drive

---

[15] American Public Transportation Association, Transit Vehicle Database, May 2006, www.apta.com/references/info/pubs

system overhauls and hybrid battery replacement for the different bus types. These assumptions on Diesel and CNG engine and transmission overhauls are based on an informal poll of maintenance staff at six transit agencies conducted by the author in 2004[10]. The assumptions for hybrid drive system overhaul, hybrid battery replacement, and fuel cell powerplant overhaul are based on discussions with system manufacturers and review of manufacturer literature.

Table 9 Overhaul Assumptions

Technology	Engine/Power plant		Transmission/Drive System		Hybrid Battery Replacement	
	Hours *	Cost	Miles	Cost	Miles**	Cost
Diesel	20,000	$17,500	100,000	$7,900	NA	NA
CNG	20,000	$22,500	100,000	$7,900	NA	NA
Diesel Hybrid	22,000	$12,500	200,000	$7,000	200,000	$30,000
Fuel Cell	10,000	$100,000	200,000	$7,000	NA	NA
Fuel Cell Hybrid	10,000	$100,000	200,000	$7,000	200,000	$30,000

* To calculate mileage interval multiply by 12.4 mph = 250,000 mi for a diesel or CNG bus and 275,000 mi for hybrid
** Nickel-metal hydride and Li-ion batteries are expected to last 6 years in a hybrid propulsion system

Given that large numbers of hybrid buses have not been in service long enough to reach expected system overhaul intervals the assumptions about hybrid drive system overhauls used in this analysis have a significant amount of uncertainty. For a series hybrid system the primary activity during hybrid drive system overhaul will be replacement of the traction motor and generator bearings. As relatively simple electric machines they should be able to go for at least twice as long as a standard automatic transmission before an overhaul is required, and bearing replacement is relatively inexpensive.

The assumed reduced cost of engine overhaul for Diesel Hybrid buses compared to Diesel buses is due to the fact that hybrid systems can use smaller and less expensive medium-duty diesel engines that would normally be installed in a pick-up truck, as opposed to the heavy-heavy duty diesel engines typically installed in Diesel transit buses.

During a Fuel Cell powerplant overhaul the major activity will be a complete replacement of the fuel cell stacks. The assumption used in this analysis of a 10,000 hour replacement interval and $100,000 replacement cost for fuel cell stacks is a forward-looking assumption.

1.2.4 Variable Overhaul Intervals (Worksheet I4)

The model used for this analysis allows the user to specify variable overhaul costs and variable overhaul intervals throughout a bus' life. For example, one could assume that as Fuel Cell technology matures fuel cell powerplant overhaul intervals will increase (i.e. fuel cell stacks will become more durable) and replacement cost will decrease, within the life time of a bus.

For this base case analysis all overhaul costs and intervals were assumed to be constant. No sub-systems for any bus type were assumed to have variable overhaul intervals or costs.

1.2.5 Depot Infrastructure Costs (Worksheet I5)

The assumptions used in this analysis for the cost of CNG fuel station installation, and depot changes required for CNG buses, is taken from the Transit Costs 1.0 model developed for the U.S. Department of Energy by TIAX, LLC[16]. This model assumes that CNG fuel stations have a fixed cost of $200,000 and a variable cost of $800 per standard cubic foot per minute (scfm) station capacity. The required scfm capacity of the station is based on the number of buses, the amount of fuel each bus will use every day, the maximum allowable fill time per bus, and the total available fueling hours per day at the bus depot. Station scfm is calculated using equations 1 and 2.

$$\# Nozzles = \frac{\# bus \times t_{fill}(\frac{min}{day})}{avail.hrs \times 60(\frac{min}{hr})} \quad \text{(equation 1)}$$

$$SCFM = \frac{\frac{miles}{yr} \times 126 \frac{scf}{DEG}}{312 \frac{day}{yr} \times \frac{miles}{DEG} \times t_{fill}(\frac{min}{day})} \times \# Nozzles \quad \text{(equation 2)}$$

Assuming 100 assigned buses, a six minute "fast fill" for each bus, and six to eight hours per day available for fueling, two CNG fueling nozzles will be required. Assuming 33,000 annual miles per bus and CNG bus fuel economy of 2.4 MPDEG, the fuel station will need to have a capacity of 1,850 scfm, rounded up to 2,000 scfm. The cost of the CNG fuel station will therefore be $1.8 million. This does not include any costs for extending natural gas lines to the location of the CNG fuel station. Depending on current installed capacity of the local natural gas utility these costs can be significant, but are unique to each facility location.

Facility design for compressed natural gas operations generally requires installation of a building methane detection system and additional building ventilation for gas purging, as required. It also requires that all potential ignition sources (including standard electrical fixtures and conduit) not be located within 18-24 inches of ceiling level, and that the building roof structural design not allow for dead pockets at ceiling level where released gas could collect without being purged by the building's ventilation system. Many existing facilities built for diesel vehicles require modifications to both HVAC and electrical systems when CNG buses are introduced.

Transit Costs 1.0 assumes that these CNG facility requirements have a fixed cost of $100,000 plus a variable cost of $2,500 per bus if buses will be stored out doors and $4,000 per bus if they will be stored in doors. This results in a cost of $350,000 - $500,000 for CNG facility modifications for a 100-bus fleet.

---

[16] Kassoy, E.; Kamakate, F.; Leonard, J.; TIAX LLC, *Transit Costs1.0;* September 2003; Developed under contract to U.S. Department of Energy; www.eere.gov/afdc/apps/toolkit/docs/Mod09b_Transitcost.xls

Diesel and Hybrid buses use diesel fuel. They require the installation of a diesel fuel storage system with dispenser(s) and do not require any other special building systems[17]. Based on the author's experience at MTA New York City Transit the cost of diesel fuel stations are generally approximately one tenth the cost of CNG fuel stations which can handle the same number of buses. This analysis therefore assumes that the cost of a diesel fuel station that can accommodate 100 buses will be $180,000.

Because hybrid systems incorporate a significant number of batteries, this analysis also assumes that the bus depot will require modifications/expansion of its existing battery room to accommodate Diesel Hybrid and Fuel Cell Hybrid buses. The assumption used for the cost of these modifications is $20,000.

The model also assumes that CNG, Diesel Hybrid, Fuel Cell, and Fuel Cell Hybrid buses will require the installation of an overhead crane at the maintenance facility, since all of these bus types usually incorporate more roof-mounted equipment than standard Diesel buses. The assumption used for the cost of this crane is $25,000.

Given the limited U.S. experience with Fuel Cell buses and hydrogen fueling infrastructure it is more difficult to determine appropriate assumptions for the cost of installing a hydrogen fuel station and modifying a depot to handle hydrogen-fueled buses. Fueling station costs will also depend on the method used for fueling.

NREL reports that VTA purchased their hydrogen fuel station, which is designed to handle a maximum of six buses, for $640,000. The VTA fuel station stores liquid hydrogen which is then vaporized and compressed onto the buses.

Sunline and AC Transit both chose to create hydrogen on site using a natural gas reformer. NREL reports that Sunline purchased, for $750,000, a commercial unit that can create and store up to 9 kg/hr of hydrogen at 5,000 psi.

Other researchers have estimated the cost of hydrogen fueling infrastructure in the context of analyses of the "transition costs" to a hydrogen economy. All of these analyses are based on conversion of privately-owned public gas stations to hydrogen operations to service a relatively small number of light-duty fuel cell cars. Their estimates range from $800,000 to over $5 million for the construction of a single hydrogen station capable of producing and dispensing between 24 kg and 3,000 kg per day or hydrogen. The analyses which evaluated the cost of both small (< 100 kg/day) and large (>1,000 kg/day) stations generally assumed large economies of scale, with the relative capital cost per unit of capacity (daily kg) falling by 50% or more as station size increased from 100 to 1,000+ kg/day.

Based on the fuel economy assumptions used in this analysis a Fuel Cell bus would consume 0.40 kg hydrogen/mile and a Fuel Cell Hybrid bus would consume 0.22 kg/mile. In this analysis all buses are assumed to travel approximately 100 miles/day, so that each Fuel Cell bus would consume 40 kg/day of hydrogen, and a fleet of 100 Fuel

---

[17] While building codes have specific requirements for facilities that will house diesel fueled vehicles, most bus facilities are, or would be, designed for the use of diesel fuel absent the introduction of natural gas or hydrogen vehicles. The cost of diesel fuel design is therefore assumed to be included in the base facility costs and the cost of CNG- and hydrogen-specific systems included in the model is for the incremental cost of designing for these operations.

Cell buses would consume 3,400 kg/day[18]. Each Fuel Cell Hybrid bus would consume 22 kg/day of hydrogen, and a fleet of 100 Fuel Cell Hybrid buses would consume 1,870 kg/day.

SUMMARY - PROJECTED HYDROGEN FUEL STATION COSTS FOR 100-BUS FLEET

SOURCE	Scaled from Actual Costs for Small Bus Demonstration Fuel Station - Assuming 50% Economies of Scale		Scaled from Projections for Small Public Light Duty Fuel Station - Assuming 50% Economies of Scale			Scaled from Projections for Medium/Large Public Light Duty Fuel Station - Assuming 0% to 85% Economies of Scale		BASE CASE RANGE OF COSTS FOR HYDROGEN FUEL STATION
	VTA	Sunline	LOW A.D. Little (2002)	HIGH A.D. Little (2002)	SFA Pacific (2002)	NREL (2005)	NREL (2006)	
Fuel Cell Buses	$5.3	$7.1	$3.1	$5.1	$5.9	$14.7	$3.9	$3.5 - $7.0
Fuel Cell Hybrid Buses	$2.9	$3.9	$1.7	$2.8	$3.2	$8.1	$2.2	$1.7 - $4.0

Fuel Station Size: 3,400 kg/day for Fuel Cell Buses and 1,870 kg/day for Fuel Cell Hybrid Buses

Sources:
- VTA — NREL/TP-560-40615
- Sunline — NREL/TP-560-41001
- A.D. Little — JT/SL 35340.020602 Fuel Choice Phase 2 Final Report
- SFA Pacific — NREL/SR-540-32525, Hydrogen Supply Cost Estimate for Hydrogen Pathways - Scoping Analysis
- NREL 2005 — NREL/CP-540-37903, Analysis of Hydrogen Infrastructure Needed to Enable Commercial Introduction of Fuel
- NREL 2006 — NREL/TP-540-38351, Hydrogen Infrastructure Transition Analysis

Table 10 Projected Hydrogen Fuel Station Costs for 100-Bus Fleet

Table 10 shows the projected capital costs of hydrogen fuel stations this large, based on the cost of the VTA and Sunline fuel stations, and based on the other published cost estimates discussed above. For each projection the published cost estimate was multiplied by a scaling factor based on the required volume (kg/day) to service 100 buses, compared to the station volume used to develop the estimate. When scaling estimates based on small stations, total costs were reduced by 50% to account for economies of scale. Based on these projected estimates, the base case assumes that a hydrogen fuel station sized to accommodate 100 Fuel Cell buses would cost $3.5 – $7.0 million, and one sized to accommodate 100 Fuel Cell Hybrid buses would cost $1.7 - $4.0 million. These assumed costs are two to four times greater than the assumed base case cost of a CNG fuel station.

The same types of modifications required at a depot to safely handle natural gas are also required to handle hydrogen. Unlike for natural gas, however, the building codes relevant to hydrogen are not well developed at this time. This has lead to a wide range of facility modification costs for the fuel cell bus demonstration projects implemented to date. For example, VTA reports spending $4.4 million on facility modifications to handle three fuel cell buses, while AC Transit reports spending $1.5 million for the same number of buses, and Sunline reports spending only $50,000 to accommodate one fuel cell bus (see Table 4). For this analysis we assumed that the cost of facility modifications to accommodate a 100-bus fleet of Fuel Cell or Fuel Cell Hybrid buses

---

[18] This calculation assumes that only 85 buses out of 100 will be in service each day.

would be double the costs to accommodate the same number of CNG buses – or $700,000 - $1,000,000.

This analysis assumes that all infrastructure investments will have a useful life of 20 years.

For all infrastructure investments (fuel station, depot modifications) this analysis assumes that the annual cost of operations and maintenance would be 5% of installed capital costs.

1.2.6 Bus Technology Training Requirements (Worksheet I6)

This analysis assumes that bus mechanics will require an average of 20 hours each of initial training on Diesel buses and five hours of annual refresher training, while bus operators will require two hours of initial training and no annual refresher training.

The analysis assumes that bus mechanics will require more training, both initial and annual, for Diesel Hybrid, CNG, Fuel Cell, and Fuel Cell Hybrid buses, due to unfamiliarity with these systems. Incremental initial and annual CNG and Fuel Cell training requirements for bus operators and managers are primarily for safety training related to natural gas and hydrogen fuel. All of the training assumptions used in the analysis are shown in Table 11.

Table 11 Assumed Training Requirements

Initial Training (hrs)	Diesel	Diesel Hybrid	CNG	Fuel Cell	FC Hybrid
Bus Mechanics	20	30	25	35	35
Bus Operators	2	3	3	3	3
Managers	0	2	2	2	2
Annual Training (hrs)					
Bus Mechanics	5	7	7	7	7
Bus Operators	0	1	1	1	1
Managers	0	0	1	1	1

Fuel Cell Bus Life Cycle Cost Model: Best Case & Future Scenario Analysis

2. Base Case Results

The following describes the results of the Base Case analysis, which uses all of the assumptions noted in Section 1. All input and output sheets from the base case analysis are included in Appendix A.

2.1 First Year Annual Costs (Worksheet O1)

Per Bus Costs:

The base case analysis shows that first-year annual operating costs for Diesel buses will range from $167,000 to $190,000 per bus, with an average of $178,988. Costs for Diesel Hybrid buses will be marginally lower (-3%), and costs for CNG buses will be marginally higher (+9%). The analysis shows that annual costs for Fuel Cell buses will average $269,832/bus (+62% compared to diesel) and annual costs for Fuel Cell Hybrid buses will average $227,601 (+36%).

Increased fuel costs account for the majority of the increase in annual costs with Fuel Cell buses compared to Diesel buses. Fuel Cell Hybrid buses have much lower annual operating costs than Fuel Cell buses due to a significant savings in fuel use and fuel costs. The Base Case results for first year annual costs are summarized in Table 12.

Table 12 Base Case Average First Year Annual Costs per Bus

			Average Cost per Bus				
			Diesel	Diesel Hybrid	CNG	Fuel Cell	Fuel Cell Hybrid
Operator Labor			$ 131,452	$ 131,452	$ 131,452	$ 131,452	$ 131,452
Annual Maintenance	Propulsion Related	Power Plant	$ 4,890	$ 5,216	$ 5,216	$ 32,600	$ 32,600
		Drive System	$ -	$ -	$ -	$ -	$ -
		Fuel System	$ -	$ -	$ -	$ -	$ -
	Non-propulsion Related		$ 13,040	$ 13,040	$ 13,040	$ 13,040	$ 13,040
	Brake Relines		$ 1,487	$ 850	$ 1,652	$ 1,749	$ 929
	Technology-Specific Cost		$ 350	$ 350	$ -	$ -	$ -
	SUB-TOTAL		*$ 19,767*	*$ 19,456*	*$ 19,908*	*$ 47,389*	*$ 46,569*
Fuel			$ 27,769	$ 21,922	$ 30,778	$ 90,991	$ 49,580
TOTAL PER BUS			**$ 178,988**	**$ 172,829**	**$ 182,138**	**$ 269,832**	**$ 227,601**

Depot Costs:

The base case analysis shows that first-year technology-specific annual operating costs for a 100-bus depot housing Diesel buses will be $14,000. Costs will increase to $24,250 if Diesel Hybrid buses will be assigned there, due to an increase in annual training costs. CNG buses will incur additional training costs as well as additional costs for fuel station O&M and incremental depot systems O&M, so that total costs will be $136,750. Depot costs for Fuel Cell buses will total $330,500 and for Fuel Cell Hybrid buses 211,500, due to even higher fuel station and incremental depot systems O&M costs. Costs are lower for Fuel Cell Hybrid buses due to fact that the required hydrogen fuel station will be smaller and less expensive, and will therefore have lower annual O&M costs.

The Base Case results for first year annual depot costs are summarized in Table 13.

Table 13 Base Case Average First Year Annual Costs per Depot

	Average Cost per Depot				
	Diesel	Diesel Hybrid	CNG	Fuel Cell	Fuel Cell Hybrid
Fuel Station O&M	$ 9,000	$ 9,000	$ 90,000	$ 262,500	$ 142,500
Incremental Depot Systems Maintenance	$ -	$ 1,000	$ 21,250	$ 42,500	$ 43,500
Maintenance of Special Tools	$ -	$ 1,250	$ 1,250	$ 1,250	$ 1,250
Maintenance of Special Infrastructure	$ -	$ -	$ -	$ -	$ -
Annual Refresher Training	$ 5,000	$ 22,000	$ 24,250	$ 24,250	$ 24,250
TOTAL FOR DEPOT	$ 14,000	$ 33,250	$ 136,750	$ 330,500	$ 211,500

2.2 Capital Costs (Worksheet O2)

The base case analysis shows that capital costs to purchase a 100-bus Diesel fleet and make technology-specific infrastructure investments total $32.93 million. With an 80% federal cost share this will require $6.59 million in local capital funds. Capital costs for the purchase of 100 CNG buses and necessary infrastructure total $40 million (+21%), while they total $50.5 million (+51%) for Diesel Hybrid buses.

Table 14 Total Capital Costs for 100-Bus Fleet and Infrastructure Investments

	Average Cost per Bus				
	Diesel	Diesel Hybrid	CNG	Fuel Cell	Fuel Cell Hybrid
Bus Purchase (mil$) (1)	$ 32.70	$ 50.20	$ 37.70	$ 320.00	$ 320.00
Fuel Station (mil$)	$ 0.18	$ 0.18	$ 1.80	$ 5.25	$ 2.85
Depot Changes ($mil)	$ -	$ 0.02	$ 0.43	$ 0.85	$ 0.87
Special Tools ($mil)	$ -	$ 0.03	$ 0.03	$ 0.03	$ 0.03
Special Infrastructure ($mil)	$ -	$ -	$ -	$ -	$ -
Initial Training ($mil)	$ 0.05	$ 0.08	$ 0.07	$ 0.08	$ 0.08
TOTAL ($mil)	$ 32.93	$ 50.50	$ 40.02	$ 326.21	$ 323.83
LOCAL SHARE	$ 6.59	$ 10.10	$ 8.00	$ 65.24	$ 64.77
FEDERAL SHARE	$ 26.34	$ 40.40	$ 32.02	$ 260.97	$ 259.06

Fuel Cell Bus Life Cycle Cost Model: Best Case & Future Scenario Analysis

Table 15 Annualized Capital Costs for 100-Bus Fleet and Infrastructure Investments

	Average Cost per Bus				
	Diesel	Diesel Hybrid	CNG	Fuel Cell	Fuel Cell Hybrid
Bus Purchase ($mil) (1)	$ 3.69	$ 5.66	$ 4.25	$ 36.10	$ 36.10
Fuel Station ($mil)	$ 0.01	$ 0.01	$ 0.14	$ 0.42	$ 0.23
Depot Changes ($mil)	$ -	$ 0.00	$ 0.03	$ 0.07	$ 0.07
Special Tools ($mil)	$ -	$ 0.00	$ 0.00	$ 0.00	$ 0.00
Special Infrastructure ($mil)	$ -	$ -	$ -	$ -	$ -
Initial Training ($mil)	$ 0.01	$ 0.01	$ 0.01	$ 0.01	$ 0.01
TOTAL ANNUALIZED	$ 3.71	$ 5.69	$ 4.44	$ 36.61	$ 36.41

The purchase of 100 Fuel Cell buses and necessary infrastructure will cost $326.2 million, almost ten times more than the purchase of Diesel buses, and will require over $65 million in local capital funding. The purchase of 100 Fuel Cell Hybrid buses and necessary infrastructure will cost several million dollars less because the required hydrogen fuel station can be smaller and therefore less expensive.

The equivalent annualized cost for this amount of capital spending is $3.71 million for Diesel buses, $4.44 million for CNG buses, $5.69 million for Diesel Hybrid buses, $36.61 million for Fuel Cell buses and $36.41 million for Fuel Cell Hybrid buses. This figure takes into account the fact that infrastructure investments have a longer useful life (20 years) than buses (12 years).

The Base Case results for total capital costs and annualized capital costs are summarized in Tables 14 and 15.

2.3 *Overhaul Costs Per Bus (Worksheet O3)*

The Base Case results for total life-time overhaul costs are summarized in Table 16. Overhauls for diesel and CNG buses include one base bus and one engine overhaul (in years seven and eight, respectively) and three transmission overhauls (in years four, seven, and ten). Hybrid bus overhauls include a base bus overhaul, drive system overhaul, and battery replacement in year seven and an engine overhaul in year nine. Both Fuel Cell and Fuel Cell Hybrid overhauls include fuel cell stack replacement in years four and eight, and a base bus and drive system overhaul in year seven. Fuel Cell Hybrid also includes a hybrid battery replacement in year seven.

Table 16 Life Time Overhaul Costs per Bus

	Life Time Overhaul Costs per Bus				
	Diesel	Diesel Hybrid	CNG	Fuel Cell	Fuel Cell Hybrid
Total	$ 105,035	$ 114,712	$ 110,898	$ 289,647	$ 324,032
NPV of Total	$ 73,962	$ 80,533	$ 77,930	$ 213,871	$ 238,309

As shown in Table 16 total overhaul costs are marginally higher for Hybrid and CNG buses than for Diesel buses. Overhaul costs for Fuel Cell and Fuel Cell Hybrid buses are approximately three times higher than for diesel buses.

2.4 Total Life Cycle Costs (Worksheet O4)

The total life cycle costs of the various bus/technology types analyzed are summarized in Figures 1 – 5. Figure 1 shows *total* life cycle costs per bus (net present value) for each bus/technology type, while Figure 2 shows the *local* life cycle costs per bus. The difference between these two figures is that local costs in Figure 2 do not include the portion of capital costs paid by the federal government.

Figure 3 shows the average total annual costs per bus (in current dollars). Figure 4 shows average total life cycle costs per mile (in current dollars) and Figure 5 shows the average local life cycle costs per mile (in current dollars). Figures 4 and 5 also includes 'error bars' showing the range of costs projected by the life cycle cost model based on the high and low values input for each cost assumption.

As shown in Figure 1 Diesel, Diesel Hybrid, and CNG buses have similar total life cycle costs of $2.2 million, $2.3 million, and $2.3 million per bus, respectively. Life time Fuel Cell bus costs are almost three times higher at $6.2 million per bus. Life time Fuel Cell Hybrid bus costs are slightly lower at $5.8 million per bus.

Figure 1 Average Total Life Cycle Costs per Bus (net present value)

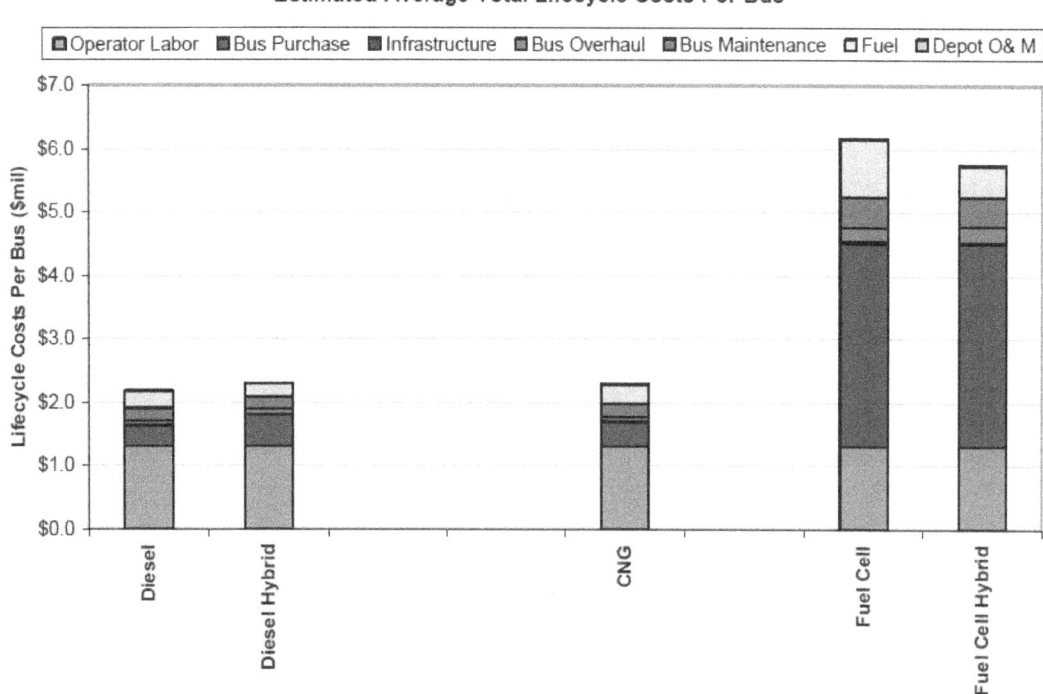

Fuel Cell Bus Life Cycle Cost Model: Best Case & Future Scenario Analysis

Figure 2 Average Local Life Cycle Costs per Bus (net present value)

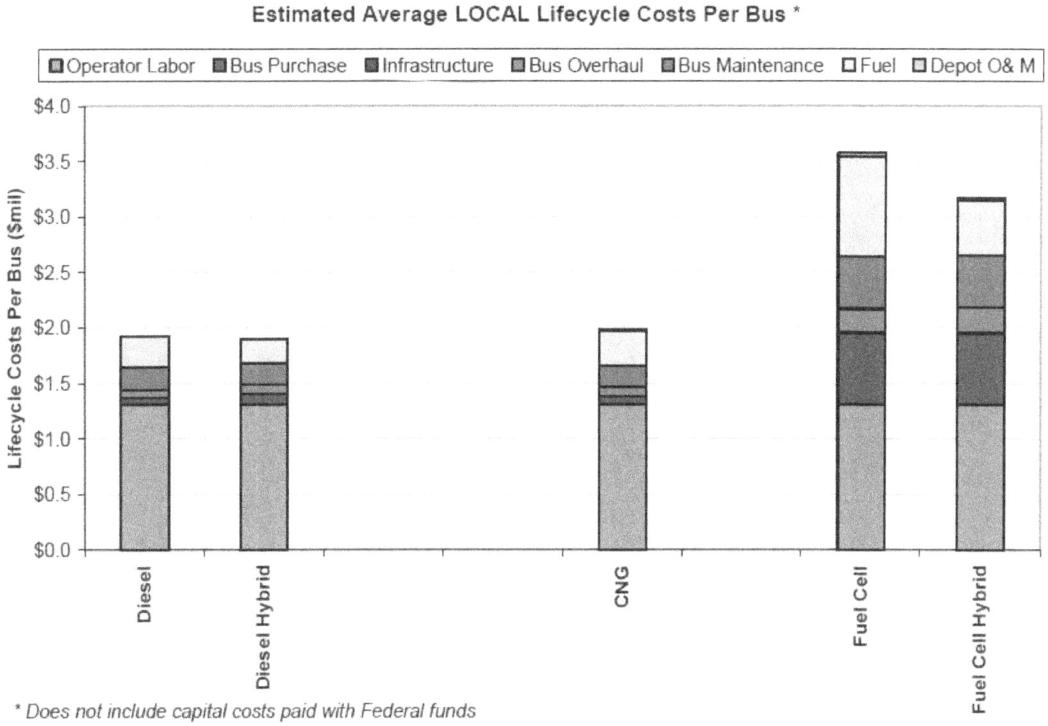

Estimated Average LOCAL Lifecycle Costs Per Bus *

[Bar chart showing lifecycle costs per bus ($mil) for Diesel (~$1.9M), Diesel Hybrid (~$1.9M), CNG (~$2.0M), Fuel Cell (~$3.6M), and Fuel Cell Hybrid (~$3.2M). Categories: Operator Labor, Bus Purchase, Infrastructure, Bus Overhaul, Bus Maintenance, Fuel, Depot O&M.]

* Does not include capital costs paid with Federal funds

As shown, the single biggest contributor to the increased life cycle costs for Fuel Cell and Fuel Cell Hybrid buses is the increased capital cost to purchase buses and install necessary infrastructure. However, all cost elements other than operator labor costs are significantly higher for fuel cell buses than for the other bus types, including life time overhaul costs (~3x higher), annual maintenance costs (~2 x higher), and fuel costs (~3x higher for Fuel Cell and ~2x higher for Fuel Cell Hybrid).

As shown in Figure 2, if only locally paid costs are included (not including the portion of capital costs paid with federal funds), life cycle costs per bus fall to $1.9 million for Diesel and Diesel Hybrid buses, $2.0 million for CNG buses, $3.6 million for Fuel Cell buses, $3.2 million for Fuel Cell Hybrid buses.

As shown in Figure 3 average annual costs for Diesel, CNG, and Diesel Hybrid buses are approximately $200,000 per bus, while they are approximately $514,000 per bus for Fuel Cell buses and $479,000 for Fuel Cell Hybrid buses.

As shown in Figure 4 total per mile costs for Diesel buses range from $5.28 to $5.89, with an average of $5.58 per mile. CNG bus costs average $5.87/mile (+5%) and Diesel Hybrid costs average $5.90/mile (+5%). Fuel Cell bus costs range from $14.97 to $16.59/mile, with an average of $15.78/mile. Fuel Cell Hybrid bus costs are slightly lower, averaging $14.70/mile, with a range of $14.09 to $15.31/mile

As shown in Figure 5, if only locally paid costs are included, per mile life cycle costs fall to $4.91 for Diesel buses, $4.86 for Diesel Hybrid Buses, $5.06 for CNG buses, $9.15 for Fuel Cell Buses, and $8.10 for Fuel Cell Hybrid buses.

Fuel Cell Bus Life Cycle Cost Model: Best Case & Future Scenario Analysis

Figure 3 Average Annual Costs per Bus (current dollars)

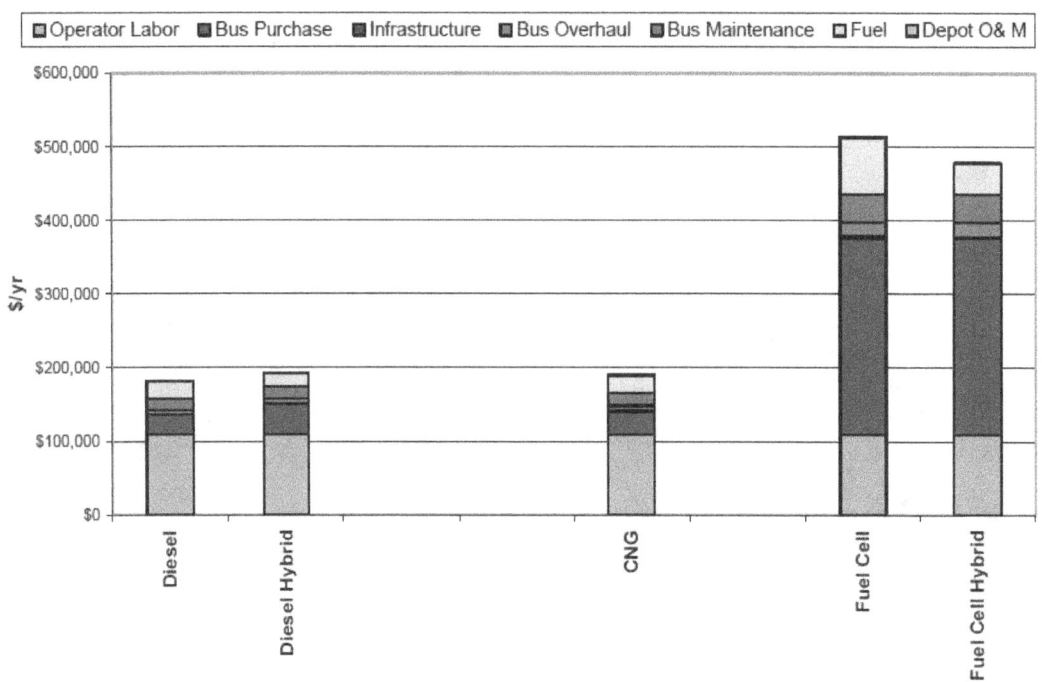

Figure 4 Average Total Life Cycle Costs per Mile (current dollars)

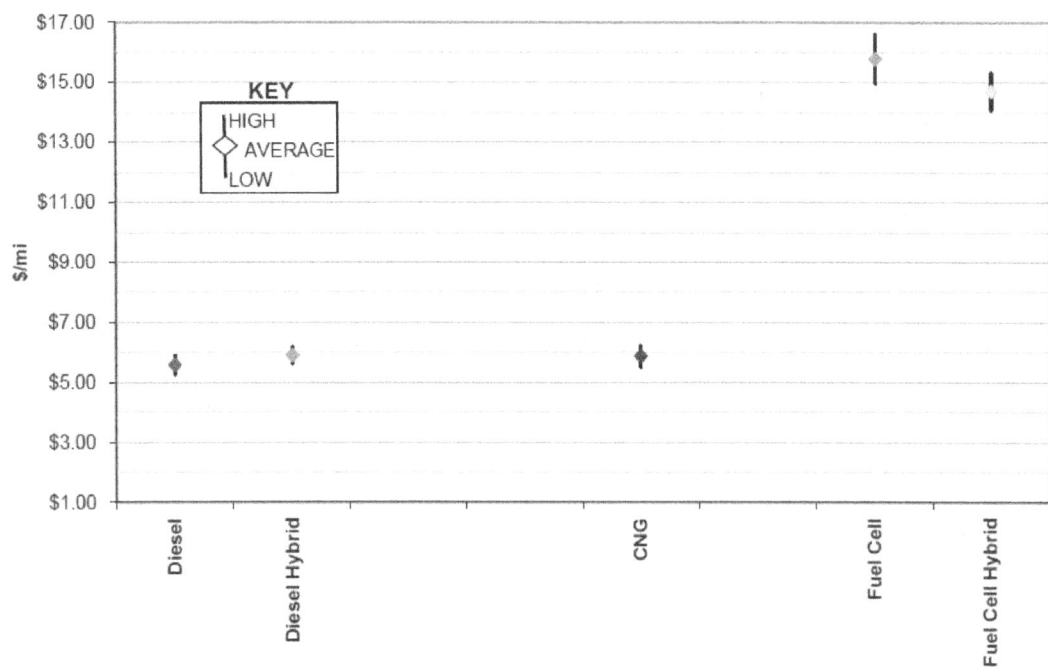

Fuel Cell Bus Life Cycle Cost Model: Best Case & Future Scenario Analysis

Figure 5   Average Local Life Cycle Costs per Mile (current dollars)

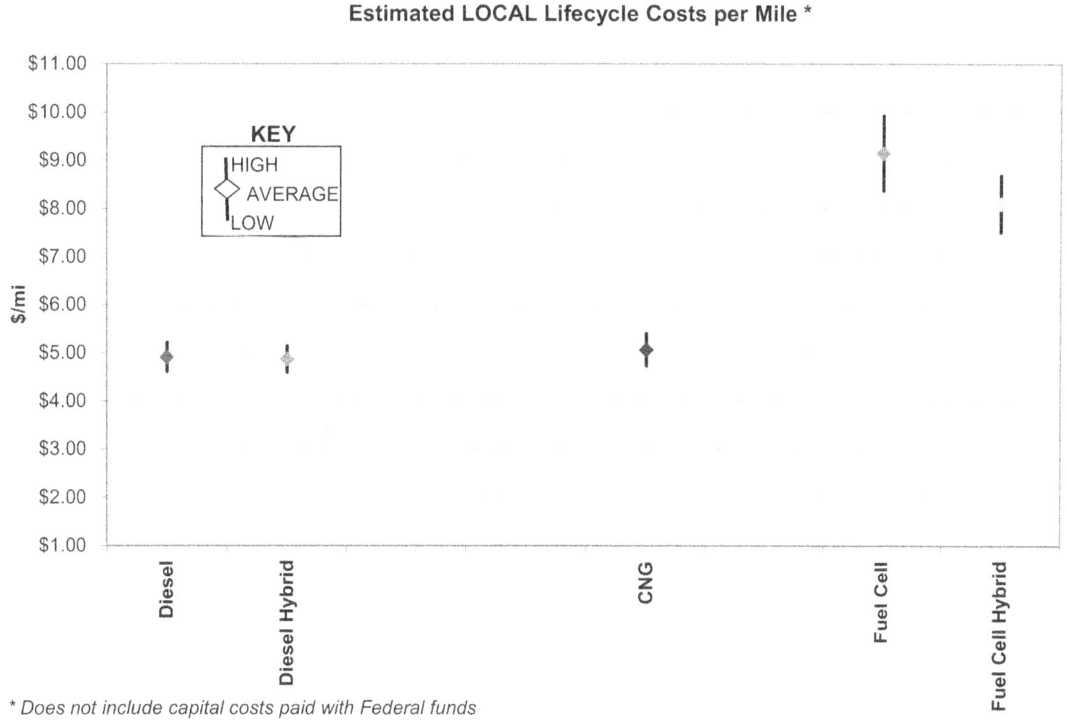

Estimated LOCAL Lifecycle Costs per Mile *

Does not include capital costs paid with Federal funds

Figures 6 and 7 show the percentage distribution of total lifecycle costs and local life cycle costs, respectively, for Diesel and Fuel Cell buses. As shown in Figure 6 operator costs make up 60% of total life cycle costs for Diesel buses; the second largest cost element is amortization of capital costs, at 15%. The distribution of costs is similar for both Diesel Hybrid and CNG buses.

With Fuel Cell buses amortization of capital costs accounts for over 50% of total life cycle costs, pushing operator costs down to only 21% of the total. Though higher in total for Fuel Cell buses, the other cost categories (overhaul costs, maintenance costs, fuel costs, and depot costs) comprise a similar percentage of the total for both Diesel and Fuel Cell buses.

As shown in Figure 7, if only locally paid costs are included operator costs account for over 68% of total costs for Diesel buses; the second highest cost category is fuel costs at 14.4%, and capital costs only account for 3.4% of local costs. By contrast, operator costs only account for 36.5% of local costs for fuel cell buses. Capital costs still account for almost 18% of local costs and fuel accounts for over 25% of local costs.

Figure 6 Percentage of Total Life Cycle Costs by Cost Category

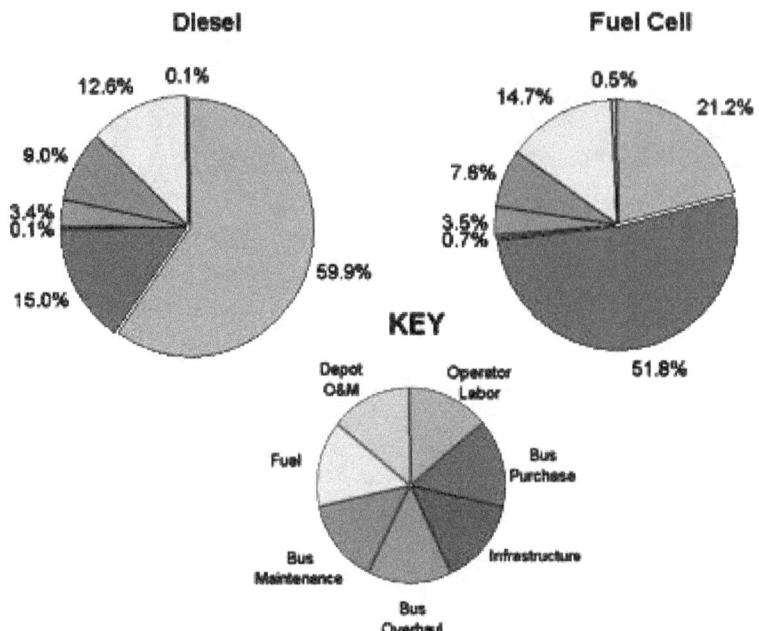

Figure 7 Percentage of Local Life Cycle Costs by Cost Category

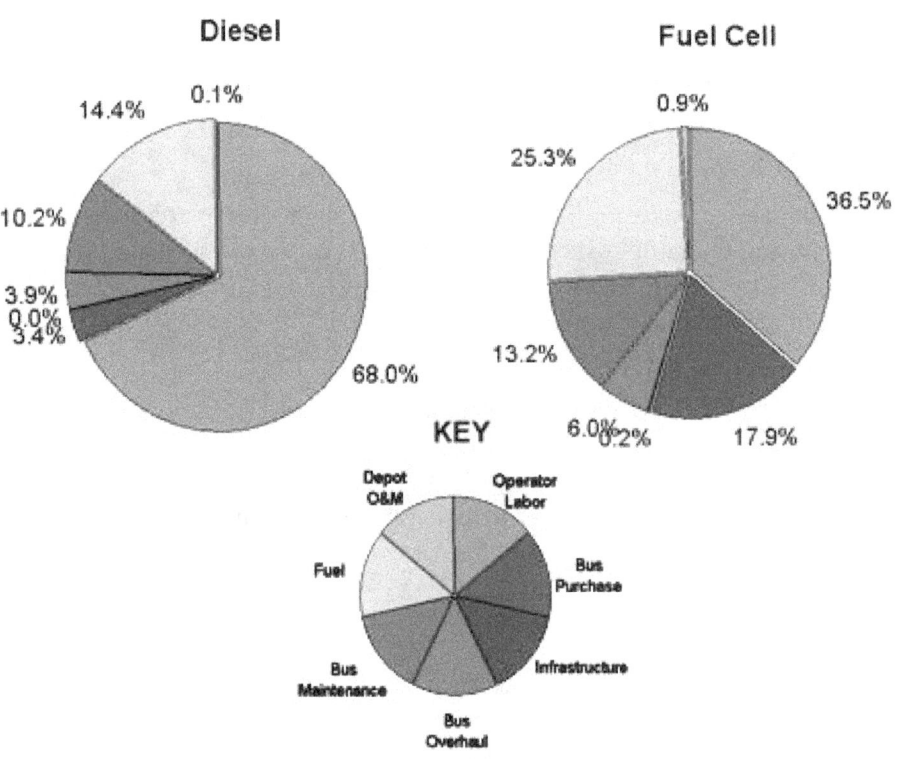

3. Future Cost Scenarios

This section describes the results of a "best case" analysis, and sensitivity analyses conducted using the life cycle cost model.

The best case scenario was intended to evaluate the potential for near-term fuel cell bus cost reductions if current federal performance goals can be met. The sensitivity analyses were intended to explore the effect on total life cycle costs of several major fuel cell bus cost drivers, including bus purchase cost and hydrogen fuel cost.

3.1 "Best Case" Assumptions

The assumptions used for the best case scenario are primarily based on meeting the Federal Transit Administration's near-term National Fuel Cell Bus performance objectives, and the U.S. Department of Energy's 2015 goal for the cost of hydrogen fuel.

The FTA fuel cell bus performance objectives include:

- Fuel Cell bus purchase cost ≤ 5x diesel bus purchase cost
- Fuel Cell stack durability of 20,000 - 30,000 hrs
- Double the fuel economy of a diesel bus

DOE's 2015 goal for the delivered cost of hydrogen is ≤ $3.00/kg (untaxed) in 2005 dollars. This is equivalent to $3.39/DEG, a greater than 50% reduction compared to the base case assumption.

In order to meet the FTA fuel economy goal, the best case scenario assumes that any fuel cell bus would need to be a Fuel Cell Hybrid bus.

Table 17 Major Best Case Assumptions

	UNIT	Best Case Fuel Cell Hybrid	Base Case Fuel Cell Hybrid	Base Case Fuel Cell
Bus Purchase	$ mill	$1.6	$3.2	$3.2
Fuel Cell Stack Life	hrs	20,000 – 30,000	10,000	10,000
Fuel Cell Stack Cost	$	$50,000	$100,000	$100,000
Fuel Economy	MPDEG	5.2 – 7.6	4.2 – 6.1	2.3 – 3.3
Hydrogen Cost	$/kg	$3.00	$6.70	$6.70
Propulsion Maintenance	$/mi	$0.20 - $0.40	$0.75 – $1.25	$0.75 – $1.25
Hydrogen Fuel Station	$ mill	$1.8	$1.7 - $4.0	$3.5 - $7.0
Hybrid Battery Cost	$	$20,000	$30,000	$30,000

While not included in the FTA and DOE goals, the best case scenario also assumes that hydrogen infrastructure and fuel cell bus maintenance costs will be reduced compared to the base case. The best case scenario assumes that $/mi propulsion maintenance costs for Fuel Cell Hybrid buses will be ≤ 2x $/mi propulsion maintenance costs for diesel buses, that hydrogen fuel station costs will be ≤ 2x the cost of a similar capacity CNG fuel station, that fuel cell stack replacement will cost one half of the base case cost, and that hybrid battery replacement will cost two thirds of the base case cost.

Table 17 shows all of the assumptions used in the best case analysis, compared to the parallel base case assumptions. All other assumptions used by the model that are not listed in Table 17 are the same for the base case and the best case.

3.2 "Best Case" Results

Under the best case scenario, total per-mile life cycle costs for Fuel Cell Hybrid buses fall by 40% compared to the base case, to $8.88/mile. If only local costs are included best case average per-mile life cycle costs for Fuel Cell Hybrid buses fall to $5.49/mile - $0.58/mile more than local life cycle costs for Diesel buses.

The results of the best case analysis are shown in Figures 8 -10.

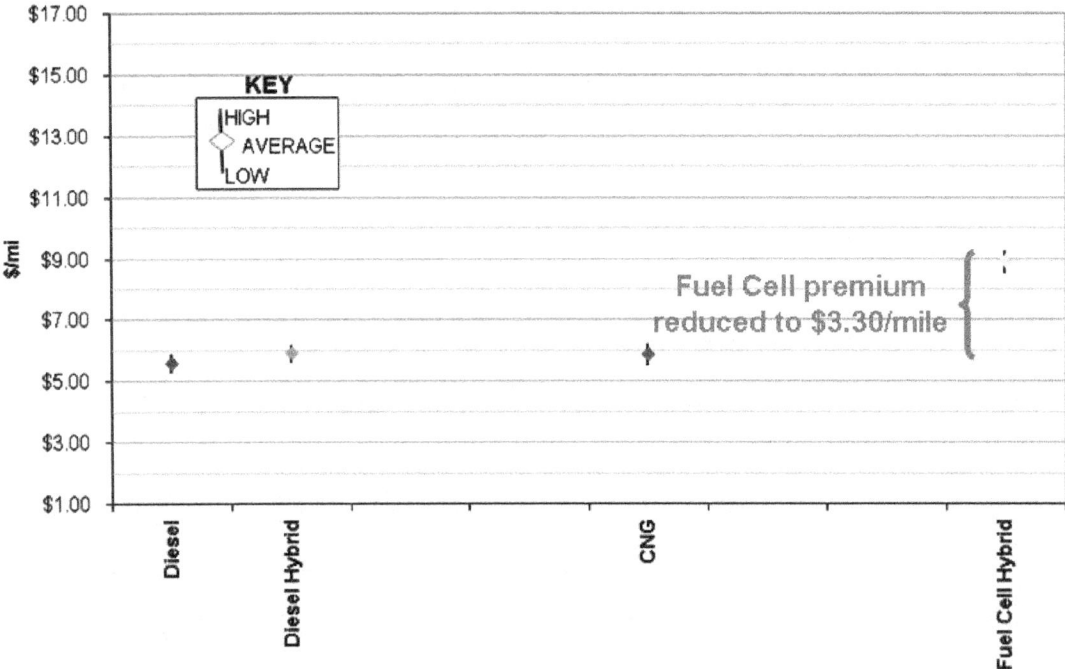

Figure 8 Best Case Total Life Cycle Costs ($/mi)

Figure 9 Best Case Local Life Cycle Costs ($/mi)

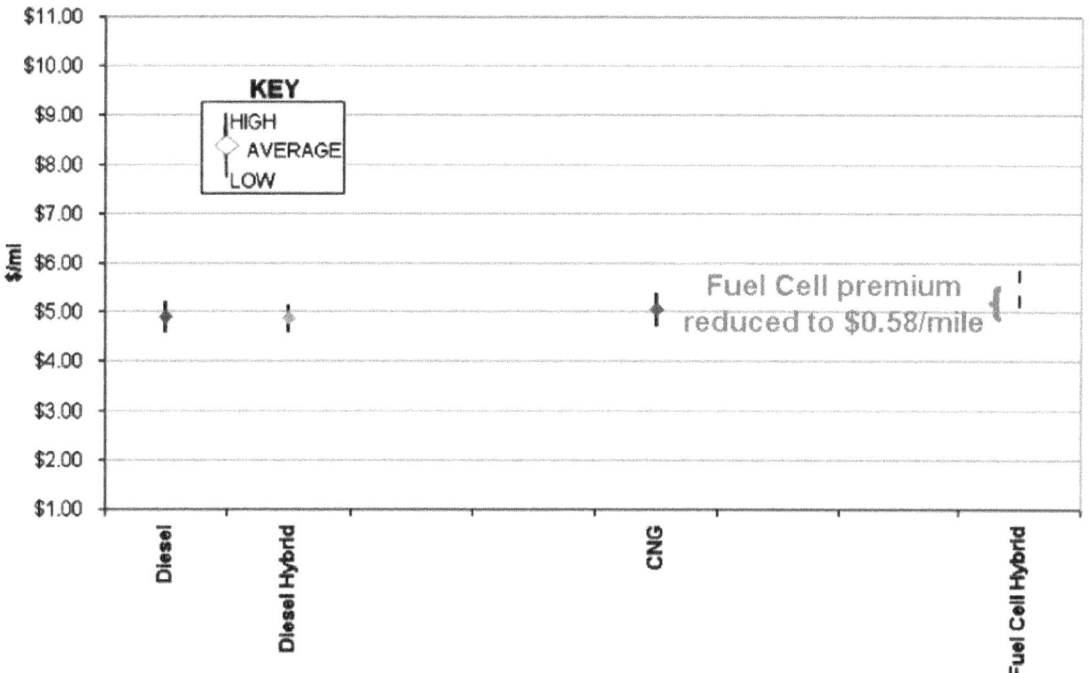

Figure 10 Best Case Total Life Cycle Costs ($/bus)

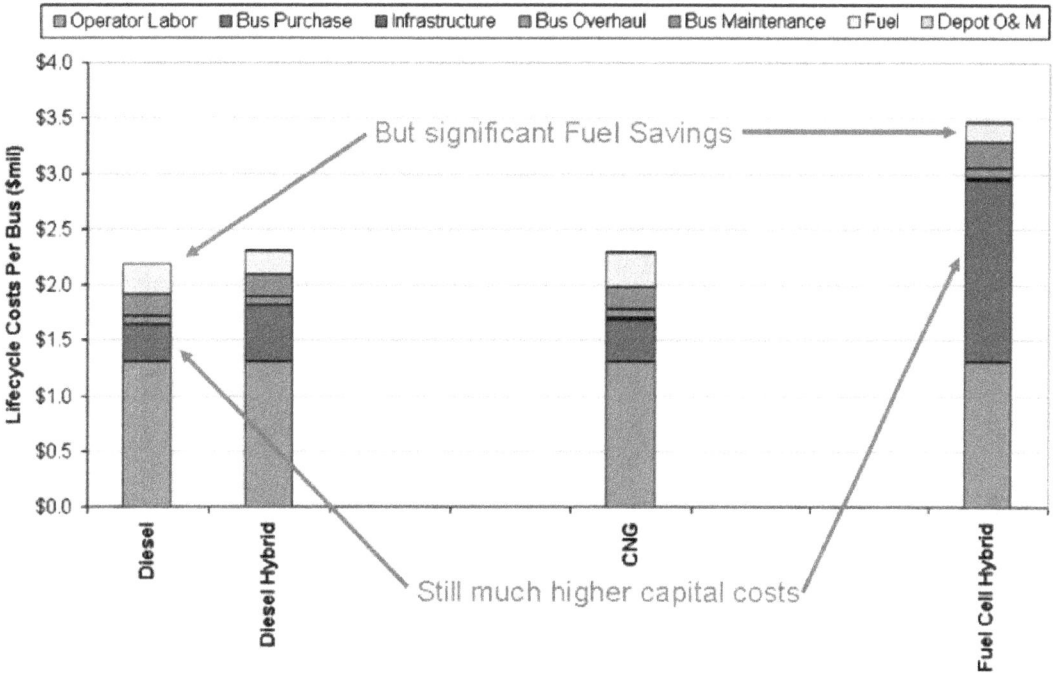

3.3 Sensitivity Analysis – Capital and Fuel Costs

Under the best case scenario the single largest contributor to higher life cycle costs for Fuel Cell Hybrid buses is still capital amortization due to a higher bus purchase price and higher infrastructure costs for hydrogen fueling. Under the best case scenario capital amortization accounts for almost 48% of total life cycle costs for Fuel Cell Hybrid buses, compared to 15% for diesel buses.

The life cycle cost model was used to evaluate the "break-even" capital cost for Fuel Cell Hybrid buses. With all other best case assumptions held constant, a Fuel Cell Hybrid bus would have to cost no more than $350,000 (less than the price of current CNG buses) for total life cycle costs to fall to the level of costs for Diesel buses. In order to match local life cycle costs for Diesel buses a Fuel Cell Hybrid bus could cost no more than $500,000 (approximately the current price of diesel hybrid buses).

Under the base case scenario all life cycle cost elements are higher for Fuel Cell and Fuel Cell Hybrid buses than for Diesel buses. Under the best case scenario, while all other cost elements are still higher, life cycle *fuel* costs are significantly lower for Fuel Cell Hybrid buses than for Diesel buses. This fuel cost savings partially off-sets the increased life cycle costs for capital amortization, maintenance, and overhauls: The lower the price of hydrogen fuel, the greater the reduction.

The life cycle cost model was used to evaluate the effect of hydrogen fuel price on total life cycle costs. This analysis is summarized in Figure 11. As shown, even if hydrogen fuel were free the fuel cost savings from Fuel Cell Hybrid buses would not fully off-set the increases in other cost categories compared to diesel buses.

Figure 11 Effect of Hydrogen Cost on Best Case Life Cycle Costs

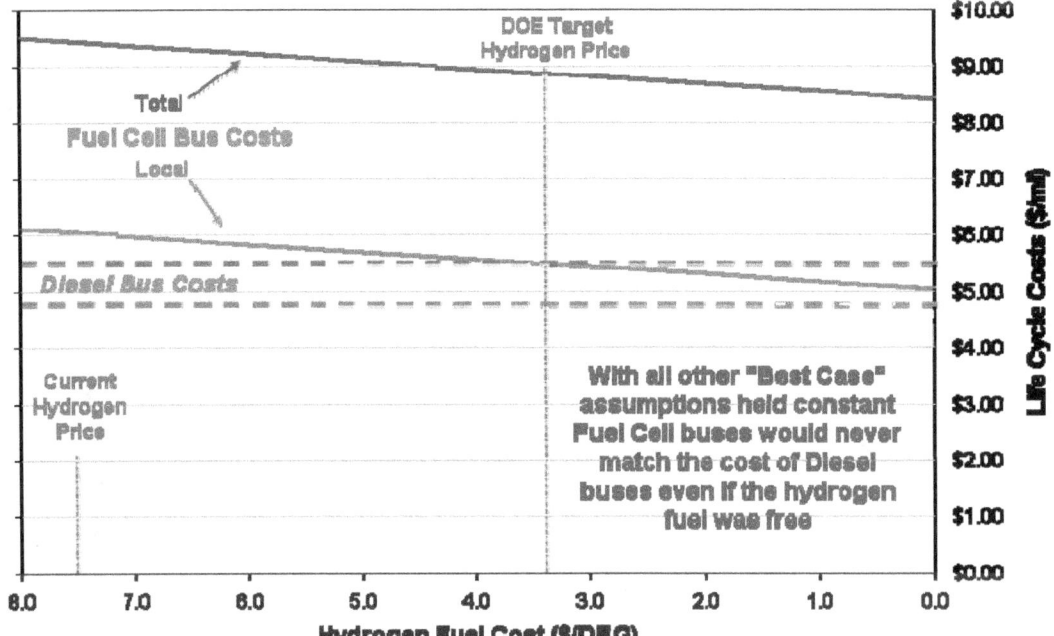

Fuel Cell Bus Life Cycle Cost Model: Best Case & Future Scenario Analysis

APPENDIX A

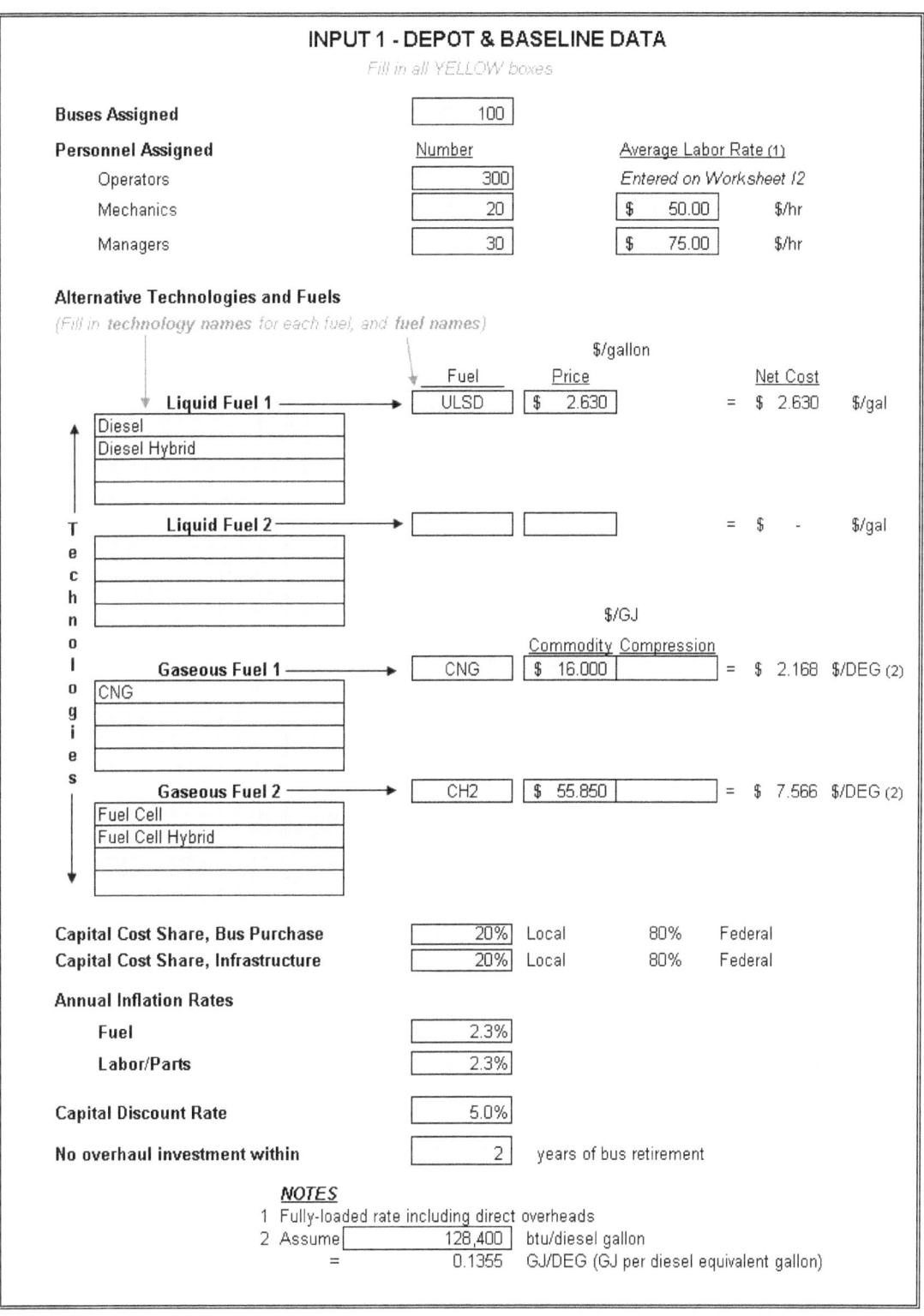

June 2007

Fuel Cell Bus Life Cycle Cost Model: Best Case & Future Scenario Analysis

APPENDIX A

INPUT 2 - ANNUAL BUS COSTS

Choose technologies, fill in all YELLOW boxes for each technology, and choose units if applicable

Choose technologies for each fuel

			Liquid Fuel 1 ULSD				Liquid Fuel 2 None				Gaseous Fuel 1 CNG				Gaseous Fuel 2 CH2			
			Diesel		Diesel Hybrid		[None]		[None]		CNG		[None]		Fuel Cell		Fuel Cell Hybrid	
		UNIT	Low	High	Low	High	Low	High	Low	High	Low	High	Low	High	Low	High	Low	High
Useful Life		yr	12		12						12				12		12	
Annual Mileage		mi/yr	32,600		32,600						32,600				32,600		32,600	
Average Speed		MPH	12.4		12.4						12.4				12.4		12.4	
Operator Labor Rate		$/hr	$ 50.00		$ 50.00						$ 50.00				$ 50.00		$ 50.00	
Annual Bus Maintenance	Propulsion Related	Power Plant	$/mi	$ 0.10	$ 0.20	$ 0.11	$ 0.21					$ 0.11	$ 0.21			$ 0.75	$ 1.25	$ 0.75	$ 1.25
		Drive System	$/mi																
		Fuel System	$/mi																
	Non-propulsion Related		$/mi	$ 0.25	$ 0.55	$ 0.25	$ 0.55					$ 0.25	$ 0.55			$ 0.25	$ 0.55	$ 0.25	$ 0.55
Bus Brake Reline		Front Interval	mi	33,000	37,000	57,750	64,750					29,700	33,300			29,050	31,450	52,800	59,200
		Rear Interval	mi	28,000	32,000	49,000	56,000					25,200	28,800			23,800	27,200	44,800	51,200
		Front Material Cost	$	$ 400.00	$ 400.00	$ 400.00	$ 400.00					$ 400.00	$ 400.00			$ 400.00	$ 400.00	$ 400.00	$ 400.00
		Rear Material Cost	$	$ 400.00	$ 400.00	$ 400.00	$ 400.00					$ 400.00	$ 400.00			$ 400.00	$ 400.00	$ 400.00	$ 400.00
		Front Labor	hr	4.0	6.0	4.0	6.0					4.0	6.0			4.0	6.0	4.0	6.0
		Rear labor	hr	7.0	9.0	7.0	9.0					7.0	9.0			7.0	9.0	7.0	9.0
Technology-Specific Costs (1)	A		$/yr	300.00	400.00	300.00	400.00												
	B		$/mi																
	C		hr/yr																
	D		$/mi																
	E		hr/yr																
Choose units																			
Bus Fuel Economy (2)		MPG	2.60	3.80	3.30	4.80					1.90	2.90			2.30	3.30	4.20	6.10
Fuel Cost (2)		$/gal	$ 2.630		$ 2.630		$ -		$ -		$ 2.168		$ 2.168		$ 7.566		$ 7.566	

NOTES

(1) *(Add details of technology specific costs) :*

A $/yr   DPF cleaning
B $/mi
C   hr/yr
D $/mi
E   hr/yr

(2) For liquid fuels, MPG and $/gallon; for gaseous fuels Miles per Diesel Equivalent gallon (MPGED), $ per diesel equivalent gallon ($/DEG)

June 2007

Fuel Cell Bus Life Cycle Cost Model: Best Case & Future Scenario Analysis

APPENDIX A

INPUT 3 - BUS PURCHASE & OVERHAUL COSTS

Fuel Technology			Liquid Fuel 1 ULSD				Liquid Fuel 2 None				Gaseous Fuel 1 CNG				Gaseous Fuel 2 CH2			
			Diesel		Diesel Hybrid		[none]		[none]		CNG		[none]		Fuel Cell		Fuel Cell Hybrid	
			Low	High	Low	High	Low	High	Low	High	Low	High	Low	High	Low	High	Low	High
Bus Purchase Cost (1)	Price	$	327,000	$ 327,000	$ 502,000	$ 502,000	$		$		$ 377,000	$ 377,000			$ 3,200,000	$ 3,200,000	$ 3,200,000	$ 3,200,000
	Credit	$																	
	Net Cost	$	327,000	$ 327,000	$ 502,000	$ 502,000	$ -	$ -	$ -	$ -	$ 377,000	$ 377,000	$ -	$ -	$ 3,200,000	$ 3,200,000	$ 3,200,000	$ 3,200,000

NOTE: Fill in Overhaul Intervals and Overhaul Costs Below ONLY if they are constant over bus life (disregarding inflation). If costs or intervals will change over bus life for any system, include in Worksheet14. Intervals can be specified in miles or hours - choose the correct metric for each.

Engine/Power Plant Overhaul (2)	Interval	hrs	20,000	20,000	22,000	22,000					20,000	20,000			10,000	10,000	10,000	10,000
	Cost	$	17,500	17,500	12,500	12,500					22,500	22,500			100,000	100,000	100,000	100,000
Transmission/Drive System Overhaul (2)	Interval	miles	100,000	100,000	200,000	200,000					100,000	100,000			200,000	200,000	200,000	200,000
	Cost	$	7,900	7,900	7,000	7,000					7,900	7,900			7,000	7,000	7,000	7,000
Bus Overhaul (2)	Interval	miles	200,000	200,000	200,000	200,000					200,000	200,000			200,000	200,000	200,000	200,000
	Cost	$	50,000	50,000	50,000	50,000					50,000	50,000			50,000	50,000	50,000	50,000
Technology Specific Overhaul A (3)	Interval	miles			200,000	200,000											200,000	200,000
	Cost	$			30,000	30,000											30,000	30,000
Technology-Specific Overhaul B (4)	Interval	miles																
	Cost	$																
Technology-Specific Overhaul C (5)	Interval	miles																
	Cost	$																

NOTES
1. Credit is any alternative fuel or other credit for the specific technology
2. Engine/Powerplant includes fuel cell stacks for fuel cell buses.
   Transmission/Drive System includes electric drive motor, inverters, power control system, and generator (if included) for hybrid and fuel cell buses
   Bus overhaul includes all other base bus systems other than engine/drive motor and transmission/drive system.
3. Add details of technology specific overhaul A: Hybrid battery replacement
4. Add details of technology specific overhaul B
4. Add details of technology specific overhaul C:

APPENDIX A

Fuel Cell Bus Life Cycle Cost Model: Best Case & Future Scenario Analysis

INPUT 4 - ANNUAL OVERHAUL COSTS IF VARIABLE

Cumulative End of Yr: Miles / Hours

For all subcategories for which overhaul costs in intervals are over the life of the bus, input TOTAL ANNUAL overhaul costs in the appropriate years, using INFLATED DOLLAR VALUES. Years in range and bus basis accumulative at the end of each year. DO NOT include on this sheet any overhaul costs for systems with defined costs and revenues on Sheet 2, rows 15-24.

Fuel	Liquid Fuel 1				Liquid Fuel 2				Gaseous Fuel 1				Gaseous Fuel 2			
Technology	ULSD		Diesel Hybrid		None		[none]		CNG		[none]		Fuel Cell		Fuel Cell Hybrid	
Year	Low $	High $	Low $	High $	Low $	High $	Low $	High $	Low $	High $	Low $	High $	Low $	High $	Low $	High $
1	30,600	2,629	30,600	2,629	retired	retired	retired	retired	2,629	32,600	retired	retired	2,629	32,600	2,629	32,600
2	65,200	5,258	65,200	5,258	retired	retired	retired	retired	5,258	65,200	retired	retired	5,258	65,200	5,258	65,200
3	97,800	7,887	97,800	7,887	retired	retired	retired	retired	7,887	97,800	retired	retired	7,887	97,800	7,887	97,800
4	130,400	10,516	130,400	10,516	retired	retired	retired	retired	10,516	130,400	retired	retired	10,516	130,400	10,516	130,400
5	163,000	13,145	163,000	13,145	retired	retired	retired	retired	13,145	163,000	retired	retired	13,145	163,000	13,145	163,000
6	196,600	15,774	196,600	15,774	retired	retired	retired	retired	15,774	196,600	retired	retired	15,774	196,600	15,774	196,600
7	228,200	18,403	228,200	18,403	retired	retired	retired	retired	18,403	228,200	retired	retired	18,403	228,200	18,403	228,200
8	260,800	21,032	260,800	21,032	retired	retired	retired	retired	21,032	260,800	retired	retired	21,032	260,800	21,032	260,800
9	293,400	23,661	293,400	23,661	retired	retired	retired	retired	23,661	293,400	retired	retired	23,661	293,400	23,661	293,400
10	326,000	26,290	326,000	26,290	retired	retired	retired	retired	26,290	326,000	retired	retired	26,290	326,000	26,290	326,000
11	358,600	28,919	358,600	28,919	retired	retired	retired	retired	28,919	358,600	retired	retired	28,919	358,600	28,919	358,600
12	391,200	31,548	391,200	31,548	retired	retired	retired	retired	31,548	391,200	retired	retired	31,548	391,200	31,548	391,200
13	retired	retired	retired	retired	retired	retired	retired	retired	retired	retired	retired	retired	retired	retired	retired	retired
14	retired	retired	retired	retired	retired	retired	retired	retired	retired	retired	retired	retired	retired	retired	retired	retired
15	retired	retired	retired	retired	retired	retired	retired	retired	retired	retired	retired	retired	retired	retired	retired	retired
16	retired	retired	retired	retired	retired	retired	retired	retired	retired	retired	retired	retired	retired	retired	retired	retired
17	retired	retired	retired	retired	retired	retired	retired	retired	retired	retired	retired	retired	retired	retired	retired	retired
18	retired	retired	retired	retired	retired	retired	retired	retired	retired	retired	retired	retired	retired	retired	retired	retired
19	retired	retired	retired	retired	retired	retired	retired	retired	retired	retired	retired	retired	retired	retired	retired	retired
20	retired	retired	retired	retired	retired	retired	retired	retired	retired	retired	retired	retired	retired	retired	retired	retired
21	retired	retired	retired	retired	retired	retired	retired	retired	retired	retired	retired	retired	retired	retired	retired	retired
22	retired	retired	retired	retired	retired	retired	retired	retired	retired	retired	retired	retired	retired	retired	retired	retired
23	retired	retired	retired	retired	retired	retired	retired	retired	retired	retired	retired	retired	retired	retired	retired	retired
24	retired	retired	retired	retired	retired	retired	retired	retired	retired	retired	retired	retired	retired	retired	retired	retired
25	retired	retired	retired	retired	retired	retired	retired	retired	retired	retired	retired	retired	retired	retired	retired	retired

June 2007

Fuel Cell Bus Life Cycle Cost Model: Best Case & Future Scenario Analysis

APPENDIX A

INPUT 5 - DEPOT INFRASTRUCTURE COSTS

For each technology, fill in purchase cost, annual maintenance costs, and useful life (yrs) for each applicable element of infrastructure. Also fill in labels for each element in cells C14-G17.

PURCHASE & INSTALLATION

	Fuel Technology	unit	Liquid Fuel 1 ULSD Diesel Low	High	Diesel Hybrid Low	High	Useful Life (yr)	Liquid Fuel 2 None [none] Low	High	[none] Low	High	Useful Life (yr)	Gaseous Fuel 1 CNG CNG Low	High	[none] Low	High	Useful Life (yr)	Gaseous Fuel 2 CH2 Fuel Cell Low	High	Fuel Cell Hybrid Low	High
Fuel Station		$ total	$ 180,000	$ 180,000	$ 180,000	$ 180,000	20					20	$ 1,800,000	$ 1,800,000			20	$ 3,500,000	$ 7,000,000	$ 1,700,000	$ 4,000,000
Depot Changes	NG Mods	$ total					20					20	$ 350,000	$ 500,000			20				
	H2 mods	$ total					0					0					20	$ 700,000	$ 1,000,000	$ 700,000	$ 1,000,000
	Battery room	$ total			$ 20,000	$ 20,000	20					0					20			$ 20,000	$ 20,000
Special Tools	Overhead crane	$ total			$ 25,000	$ 25,000	20					0	$ 25,000	$ 25,000			20	$ 25,000	$ 25,000	$ 25,000	$ 25,000
		$ total					20														
Special Infrastructure		$ total					0										0				

ANNUAL MAINTENANCE & OPERATIONS

		Unit	Liquid Fuel 1 ULSD Diesel Low	High	Diesel Hybrid Low	High	Liquid Fuel 2 0 [none] Low	High	[none] Low	High	Gaseous Fuel 1 CNG CNG Low	High	[none] Low	High	Gaseous Fuel 2 CH2 Fuel Cell Low	High	Fuel Cell Hybrid Low	High
Fuel Station		$ annual	$ 9,000.00	$ 9,000.00	$ 9,000.00	$ 9,000.00					$ 90,000.00	$ 90,000.00			$ 175,000.00	$ 350,000.00	$ 85,000.00	$ 200,000.00
Depot Systems	NG Mods	$ annual									$ 17,500	$ 25,000						
	H2 mods	$ annual													$ 35,000	$ 50,000	$ 35,000	$ 50,000
	Battery room	$ annual			$ 1,000	$ 1,000											$ 1,000	$ 1,000
Special Tools	Overhead crane	$ annual			$ 1,250	$ 1,250					$ 1,250	$ 1,250			$ 1,250	$ 1,250	$ 1,250	$ 1,250
		$ annual																
Special Infrastructure		$ annual																

June 2007

Fuel Cell Bus Life Cycle Cost Model: Best Case & Future Scenario Analysis

APPENDIX A

INPUT 6 - TRAINING COSTS

Initial Training	Fuel	Liquid Fuel 1 ULSD				Liquid Fuel 2 None				Gaseous Fuel 1 CNG				Gaseous Fuel 2 CH2			
	Technology	Diesel		Diesel Hybrid		[none]		[none]		CNG		[none]		Fuel Cell		Fuel Cell Hybrid	
	unit	Low	High	Low	High	Low	High	Low	High	Low	High	Low	High	Low	High	Low	High
Bus Mechanics	hrs	18.0	22.0	25.0	35.0					20.0	30.0			30.0	40.0	30.0	40.0
Bus Operators	hrs	2.0	2.0	3.0	3.0					3.0	3.0			3.0	3.0	3.0	3.0
Managers	hrs	0.0	0.0	2.0	2.0					2.0	2.0			2.0	2.0	2.0	2.0

Annual Refresher Training	Fuel	Standard Diesel Fuel ULSD				Alternative Diesel Fuel None				Gaseous Fuel 1 CNG				Gaseous Fuel 2 CH2			
	Technology	Diesel		Diesel Hybrid		[none]		[none]		CNG		[none]		Fuel Cell		Fuel Cell Hybrid	
	unit	Low	High	Low	High	Low	High	Low	High	Low	High	Low	High	Low	High	Low	High
Bus Mechanics	hrs	5.0	5.0	7.0	7.0					7.0	7.0			7.0	7.0	7.0	7.0
Bus Operators	hrs	0.0	0.0	1.0	1.0					1.0	1.0			1.0	1.0	1.0	1.0
Managers	hrs	0.0	0.0	0.0	0.0					1.0	1.0			1.0	1.0	1.0	1.0

Fuel Cell Bus Life Cycle Cost Model: Best Case & Future Scenario Analysis

APPENDIX A

OUTPUT 1 - FIRST YEAR ANNUAL COSTS

FIRST YEAR ANNUAL PER BUS COSTS

		Liquid Fuel 1						Liquid Fuel 2			None			Gaseous Fuel 1						Gaseous Fuel 2					
		Diesel			Diesel Hybrid			ULSD						CNG			CNG			Fuel Cell			Fuel Cell Hybrid		
		Low	Average	High	Low	Average	High	Low	Average	High	Low	Average	High	Low	Average	High	Low	Average	High	Low	Average	High	Low	Average	High
Operator Labor		$131,452	$131,452	$131,452	$131,452	$131,452	$131,452							$131,452	$131,452	$131,452				$131,452	$131,452	$131,452	$131,452	$131,452	$131,452
	Power Plant	$3,260	4,890	$6,520	3,586	5,216	6,846							3,586	5,216	6,846				$24,450	32,600	40,750	$24,450	32,600	40,750
	Propulsion Drive System Related																									
	Fuel System																									
Annual Maintenance	Non-propulsion Related	8,150	13,040	17,930	8,150	13,040	17,930							8,150	13,040	17,930				8,150	13,040	17,930	8,150	13,040	17,930
	Brake Relines	1,293	1,487	1,681	739	850	961							1,436	1,652	1,868				1,521	1,749	1,978	808	929	1,051
	Technology-Specific Cost	300	350	400	300	350	400																			
	SUB-TOTAL	$13,003	$19,767	$26,531	$12,775	$19,456	$26,137							$13,172	$19,908	$26,644				$34,121	$47,389	$60,658	$33,408	$46,569	$59,731
Fuel	Commodity Cost	22,563	27,769	32,976	17,862	21,922	25,981							24,366	30,778	37,190				74,742	90,991	107,239	40,434	49,580	58,726
	Compression Cost																									
	SUB-TOTAL	$22,563	$27,769	$32,976	$17,862	$21,922	$25,981							$24,366	$30,778	$37,190				$74,742	$90,991	$107,239	$48,434	$49,580	$58,726
TOTAL PER BUS		$167,017	$178,963	$190,959	$162,088	$172,829	$183,569							$168,990	$182,738	$195,285				$240,315	$269,832	$299,349	$205,294	$227,601	$249,909

FIRST YEAR ANNUAL DEPOT COSTS

	Liquid Fuel 1						Liquid Fuel 2			Gaseous Fuel 1						Gaseous Fuel 2					
	Diesel			Diesel Hybrid						CNG			CNG			Fuel Cell			Fuel Cell Hybrid		
	Low	Average	High	Low	Average	High	Low	Average	High	Low	Average	High	Low	Average	High	Low	Average	High	Low	Average	High
Fuel Station O&M	$9,000	9,000	$9,000	$9,000	9,000	$9,000				$90,000	90,000	$90,000				$175,000	262,500	$350,000	$85,000	142,500	$200,000
Incremental Depot Systems O&M				1,000	1,000	1,000				17,500	21,250	25,000				35,000	42,500	50,000	36,000	43,500	51,000
Special Tools O&M				1,250	1,250	1,250				1,250	1,250	1,250				1,250	1,250	1,250	1,250	1,250	1,250
Special Infrastructure O&M																					
Annual Refresher Training	5,000	5,000	5,000	22,000	22,000	22,000				24,250	24,250	24,250				24,250	24,250	24,250	24,250	24,250	24,250
TOTAL FOR DEPOT	$14,000	14,000	$14,000	$33,250	33,250	$33,250				$133,000	$136,750	$140,500				$235,500	$330,500	$425,500	$146,500	$211,500	$276,500

June 2007

A7

APPENDIX A

Fuel Cell Bus Life Cycle Cost Model: Best Case & Future Scenario Analysis

OUTPUT 2 - CAPITAL COSTS

TOTAL COST	Liquid Fuel 1												Liquid Fuel 2 None						Gaseous Fuel 1						Gaseous Fuel 2						
	Diesel			ULSD			Diesel Hybrid												CNG			Gaseous Fuel 1			Fuel Cell			Gaseous Fuel 2 CH2			Fuel Cell Hybrid		
	Low	Average	High	Low	Average	High	Low	Average	High	Low	Average	High	Low	Average	High	Low	Average	High	Low	Average	High	Low	Average	High	Low	Average	High	Low	Average	High
Bus Purchase (mil$) (t)	$ 32.70	$ 32.70	$ 32.70	$ 50.20	$ 50.20	$ 50.20										$ 37.70	$ 37.70	$ 37.70	$ 320.00	$ 320.00	$ 320.00	$ 320.00	$ 320.00	$ 320.00
Fuel Station (mil$)	$ 0.18	$ 0.18	$ 0.18	$ 0.18	$ 0.18	$ 0.18										$ 1.80	$ 1.80	$ 1.80	$ 3.60	$ 4.26	$ 2.00	$ 1.71	$ 2.98	$ 4.00
Depot Changes ($mil)				$ 0.02	$ 0.02	$ 0.02										$ 0.35	$ 0.43	$ 0.50	$ 0.70	$ 0.92	$ 1.00	$ 0.72	$ 0.87	$ 1.02
Special Tools ($mil)				$ 0.03	$ 0.03	$ 0.03										$ 0.03	$ 0.03	$ 0.03	$ 0.03	$ 0.03	$ 0.03	$ 0.03	$ 0.03	$ 0.03
Special Infrastructure ($mil)																								
Initial Training ($mil)	0.05	0.05	0.05	$ 0.07	$ 0.07	$ 0.08										$ 0.07	$ 0.07	$ 0.08	$ 0.08	$ 0.08	$ 0.09	$ 0.08	$ 0.08	0.09
TOTAL ($mil)	$ 32.93	$ 32.93	$ 32.93	$ 50.50	$ 50.50	$ 50.51										$ 39.94	$ 80.07	$ 40.10	$ 324.30	$ 326.71	$ 328.11	$ 327.52	$ 329.83	$ 325.13
LOCAL SHARE	$ 6.59	$ 6.59	$ 6.59	$ 10.10	$ 10.10	$ 10.10										$ 7.99	$ 8.00	$ 8.02	$ 64.86	$ 65.34	$ 65.62	$ 64.50	$ 64.17	$ 65.03
FEDERAL SHARE	$ 26.34	$ 26.34	$ 26.35	$ 40.40	$ 40.40	$ 40.41										$ 31.96	$ 32.02	$ 32.08	$ 259.44	$ 260.97	$ 262.49	$ 253.02	$ 258.06	$ 260.11

ANNUALIZED COST (2)	Liquid Fuel 1 ULSD												Liquid Fuel 2 None						Gaseous Fuel 1 CNG						Gaseous Fuel 2 CNG						
	Diesel			Diesel Hybrid												CNG			Fuel Cell			Fuel Cell Hybrid								
	Low	Average	High	Low	Average	High	Low	Average	High	Low	Average	High	Low	Average	High	Low	Average	High	Low	Average	High	Low	Average	High
Bus Purchase ($mil) (t)	$ 3.689	$ 3.689	$ 3.689	$ 5.664	$ 5.664	$ 5.664										$ 4.254	$ 4.254	$ 4.254	$ 36.104	$ 36.104	$ 36.104	$ 36.104	$ 36.104	$ 36.104
Fuel Station ($mil)	$ 0.014	$ 0.014	$ 0.014	$ 0.014	$ 0.014	$ 0.014										$ 0.144	$ 0.144	$ 0.144	$ 0.203	$ 0.421	$ 0.562	$ 0.136	$ 0.229	0.321
Depot Changes ($mil)				$ 0.002	$ 0.002	$ 0.002										$ 0.028	$ 0.034	$ 0.040	$ 0.056	$ 0.069	$ 0.081	$ 0.068	$ 0.070	0.082
Special Tools ($mil)				$ 0.002	$ 0.002	$ 0.002										$ 0.002	$ 0.002	$ 0.002	$ 0.002	$ 0.002	$ 0.002	$ 0.002	$ 0.002	$ 0.002
Special Infrastructure ($mil)																								
Initial Training ($mil)	$ 0.005	$ 0.005	$ 0.006	$ 0.008	$ 0.008	$ 0.008										$ 0.008	$ 0.008	$ 0.008	$ 0.009	$ 0.010	$ 0.010	$ 0.009	$ 0.010	$ 0.010
TOTAL ANNUALIZED	$ 3.71	$ 3.71	$ 3.71	$ 5.69	$ 5.69	$ 5.69										$ 4.44	$ 4.44	$ 4.45	$ 36.45	$ 36.61	$ 36.76	$ 36.31	$ 36.41	$ 36.52

APPENDIX A

Fuel Cell Bus Life Cycle Cost Model: Best Case & Future Scenario Analysis

OUTPUT 3 - OVERHAUL COSTS PER BUS

YEAR	Liquid Fuel 1 Diesel Low	Liquid Fuel 1 Diesel High	ULSD Diesel Hybrid Low	ULSD Diesel Hybrid High	Liquid Fuel 2 None Low	Liquid Fuel 2 None High	Gaseous Fuel 1 CNG Low	Gaseous Fuel 1 CNG High	CNG Low	CNG High	Gaseous Fuel 2 Fuel Cell Low	Gaseous Fuel 2 Fuel Cell High	CH2 Fuel Cell Hybrid Low	CH2 Fuel Cell Hybrid High
1	$ -	$ -	$ -	$ -	retired	retired	$ -	$ -	retired	retired	$ -	$ -	$ -	$ -
2	$ -	$ -	$ -	$ -	retired	retired	$ -	$ -	retired	retired	$ -	$ -	$ -	$ -
3	$ -	$ -	$ -	$ -	retired	retired	$ -	$ -	retired	retired	$ -	$ -	$ -	$ -
4	$ 8,458	$ 8,458	$ -	$ -	retired	retired	$ 8,458	$ 8,458	retired	retired	$ 107,060	$ 107,060	$ 107,060	$ 107,060
5	$ -	$ -	$ -	$ -	retired	retired	$ -	$ -	retired	retired	$ -	$ -	$ -	$ -
6	$ -	$ -	$ -	$ -	retired	retired	$ -	$ -	retired	retired	$ -	$ -	$ -	$ -
7	$ 66,364	$ 66,364	$ 99,718	$ 99,718	retired	retired	$ 66,364	$ 66,364	retired	retired	$ 65,332	$ 65,332	$ 99,718	$ 99,718
8	$ 20,520	$ 20,520	$ -	$ -	retired	retired	$ 26,382	$ 26,382	retired	retired	$ 117,254	$ 117,254	$ 117,254	$ 117,254
9	$ -	$ -	$ 14,994	$ 14,994	retired	retired	$ -	$ -	retired	retired	$ -	$ -	$ -	$ -
10	$ 9,694	$ 9,694	$ -	$ -	retired	retired	$ 9,694	$ 9,694	retired	retired	$ -	$ -	$ -	$ -
11	Phase Out	Phase Out	Phase Out	Phase Out	retired	retired	Phase Out	Phase Out	retired	retired	Phase Out	Phase Out	Phase Out	Phase Out
12	Phase Out	Phase Out	Phase Out	Phase Out	retired	retired	Phase Out	Phase Out	retired	retired	Phase Out	Phase Out	Phase Out	Phase Out
13	retired	retired	retired	retired	retired	retired	retired	retired	retired	retired	retired	retired	retired	retired
14	retired	retired	retired	retired	retired	retired	retired	retired	retired	retired	retired	retired	retired	retired
15	retired	retired	retired	retired	retired	retired	retired	retired	retired	retired	retired	retired	retired	retired
16	retired	retired	retired	retired	retired	retired	retired	retired	retired	retired	retired	retired	retired	retired
17	retired	retired	retired	retired	retired	retired	retired	retired	retired	retired	retired	retired	retired	retired
18	retired	retired	retired	retired	retired	retired	retired	retired	retired	retired	retired	retired	retired	retired
19	retired	retired	retired	retired	retired	retired	retired	retired	retired	retired	retired	retired	retired	retired
20	retired	retired	retired	retired	retired	retired	retired	retired	retired	retired	retired	retired	retired	retired
21	retired	retired	retired	retired	retired	retired	retired	retired	retired	retired	retired	retired	retired	retired
22	retired	retired	retired	retired	retired	retired	retired	retired	retired	retired	retired	retired	retired	retired
23	retired	retired	retired	retired	retired	retired	retired	retired	retired	retired	retired	retired	retired	retired
24	retired	retired	retired	retired	retired	retired	retired	retired	retired	retired	retired	retired	retired	retired
25	retired	retired	retired	retired	retired	retired	retired	retired	retired	retired	retired	retired	retired	retired
TOTAL	$ 105,035	$ 105,035	$ 114,712	$ 114,712	$ -	$ -	$ 110,898	$ 110,898	$ -	$ -	$ 289,647	$ 289,647	$ 324,032	$ 324,032
NPV TOTAL	$ 73,962	$ 73,962	$ 80,533	$ 80,533	$ -	$ -	$ 77,930	$ 77,930	$ -	$ -	$ 213,871	$ 213,871	$ 238,309	$ 238,309

June 2007

Fuel Cell Bus Life Cycle Cost Model: Best Case & Future Scenario Analysis

APPENDIX A

OUTPUT 4 - TOTAL LIFE CYCLE COSTS

TOTAL COSTS $ Millions		Liquid Fuel 1						Liquid Fuel 2			None			Gaseous Fuel 1						Gaseous Fuel 2					
		Diesel			ULSD			Diesel Hybrid						CNG			CNG			Fuel Cell			CH2 Fuel Cell Hybrid		
	Low	Average	High	Low	Average	High	Low	Average	High	Low	Average	High	Low	Average	High	Low	Average	High	Low	Average	High	Low	Average	High
NPV of Annualized Capital Costs (1)	$ 32.9	$ 32.9	$ 32.9	$ 50.4	$ 50.4	$ 50.4							$ 39.3	$ 39.4	$ 39.4				$ 323.1	$ 324.4	$ 325.8	$ 321.8	$ 322.7	$ 323.7
NPV of Bus Overhaul Costs (1)	$ 7.4	$ 7.4	$ 7.4	$ 8.1	$ 8.1	$ 8.1							$ 7.8	$ 7.8	$ 7.8				$ 21.4	$ 21.4	$ 21.4	$ 23.8	$ 23.8	$ 23.8
NPV of Annual Operator Labor Costs (1)	$ 130.7	$ 130.7	$ 130.7	$ 130.7	$ 130.7	$ 130.7							$ 130.7	$ 130.7	$ 130.7				$ 130.7	$ 130.7	$ 130.7	$ 130.7	$ 130.7	$ 130.7
NPV of Annual Bus Maintenance Costs (1)	$ 12.9	$ 19.7	$ 26.4	$ 12.7	$ 19.3	$ 26.0							$ 13.1	$ 19.8	$ 26.5				$ 33.9	$ 47.1	$ 60.3	$ 33.2	$ 46.3	$ 59.4
NPV of Annual Bus Fuel Costs (1)	$ 22.4	$ 27.6	$ 32.8	$ 17.8	$ 21.8	$ 25.8							$ 24.2	$ 30.6	$ 37.0				$ 74.3	$ 90.5	$ 106.6	$ 40.2	$ 49.3	$ 58.4
NPV of Annual Depot Costs (1)	$ 0.1	$ 0.1	$ 0.1	$ 0.3	$ 0.3	$ 0.3							$ 1.3	$ 1.4	$ 1.4				$ 2.3	$ 3.3	$ 4.2	$ 1.5	$ 2.1	$ 2.7
NPV of TOTAL COSTS — DEPOT FLEET	$ 206.5	$ 218.4	$ 230.3	$ 220.0	$ 230.7	$ 241.4							$ 216.5	$ 229.6	$ 242.8				$ 585.8	$ 617.4	$ 649.1	$ 551.2	$ 575.0	$ 598.7
PER BUS (2)	$ 2.06	$ 2.18	$ 2.30	$ 2.20	$ 2.31	$ 2.41							$ 2.16	$ 2.30	$ 2.43				$ 5.86	$ 6.17	$ 6.49	$ 5.51	$ 5.75	$ 5.99
AVG ANNUAL PER BUS ($)	$ 172,065	$ 181,986	$ 191,906	$ 183,320	$ 192,224	$ 201,128							$ 180,383	$ 191,357	$ 202,331				$ 488,134	$ 514,508	$ 540,883	$ 459,360	$ 479,157	$ 498,953
PER MILE ($)	$ 5.28	$ 5.58	$ 5.89	$ 5.62	$ 5.90	$ 6.17							$ 5.53	$ 5.87	$ 6.21				$ 14.97	$ 15.78	$ 16.59	$ 14.09	$ 14.70	$ 15.31

LOCAL COSTS ONLY (3)
$ Millions

	Low	Average	High	Low	Average	High	Low	Average	High	Low	Average	High	Low	Average	High	Low	Average	High	Low	Average	High	Low	Average	High
NPV of LOCAL COSTS — DEPOT FLEET	$ 180.2	$ 192.1	$ 204.0	$ 179.6	$ 190.3	$ 201.0							$ 185.0	$ 198.1	$ 211.3				$ 327.3	$ 357.9	$ 388.4	$ 293.8	$ 316.8	$ 339.8
PER BUS (2)	$ 1.80	$ 1.92	$ 2.04	$ 1.80	$ 1.90	$ 2.01							$ 1.85	$ 1.98	$ 2.11				$ 3.27	$ 3.58	$ 3.88	$ 2.94	$ 3.17	$ 3.40
AVG ANNUAL PER BUS ($)	$ 150,148	$ 160,067	$ 169,986	$ 149,697	$ 158,598	$ 167,498							$ 154,172	$ 165,107	$ 176,042				$ 272,744	$ 298,215	$ 323,685	$ 244,814	$ 263,991	$ 283,168
PER MILE ($)	$ 4.61	$ 4.91	$ 5.21	$ 4.59	$ 4.86	$ 5.14							$ 4.73	$ 5.06	$ 5.40				$ 8.37	$ 9.15	$ 9.93	$ 7.51	$ 8.10	$ 8.69

NOTES
1. For 100 buses and infrastucture investments over the useful life of the buses
2. May be over different numbers of years for each bus type, depending on defined useful life.
3. Does not include capital costs paid with Federal funds. See worksheet O2.

www.ingramcontent.com/pod-product-compliance
Lightning Source LLC
Chambersburg PA
CBHW081758170526
45167CB00008B/3236